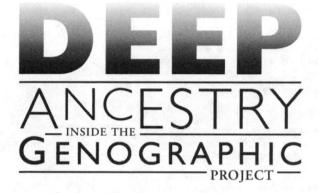

DEEP
ANCESTRY
INSIDE THE
GENOGRAPHIC
PROJECT

SPENCER WELLS

 NATIONAL GEOGRAPHIC

WASHINGTON, D.C.

ISBN-10: 0-7922-6215-8
ISBN-13: 978-0-7922-6215-2

Library of Congress Cataloging-in-Publication Data

Wells, Spencer, 1969-
  Deep ancestry : inside the Genographic Project : the landmark DNA quest to decipher
our distant past / by Spencer Wells.
    p. cm.
  Includes bibliographical references and index.
  ISBN 0-7922-6215-8 (cloth : alk. paper)
  1. Genographic Project. 2. Human evolution. 3. DNA--Evolution. 4. Human genetics.
I. Title.
QH371W45 2006
599.93'5--dc22

                               2006021421

Founded in 1888, the National Geographic Society is one of the largest nonprofit scientific and educational organizations in the world. It reaches more than 285 million people worldwide each month through its official journal, NATIONAL GEOGRAPHIC, and its four other magazines; the National Geographic Channel; television documentaries; radio programs; films; books; videos and DVDs; maps; and interactive media. National Geographic has funded more than 8,000 scientific research projects and supports an education program combating geographic illiteracy.

For more information, please call 1-800-NGS LINE (647-5463)
or write to the following address:

National Geographic Society
1145 17th Street N.W.
Washington, D.C. 20036-4688 U.S.A.

Visit us online at www.nationalgeographic.com/books

For information about special discounts for bulk purchases, please contact
National Geographic Books Special Sales: ngspecsales@ngs.org

Printed in U.S.A.

To Kim McKay,

for asking the right question

CONTENTS

INTRODUCTION

On June 26, 2000, two geneticists stood with President Bill Clinton in the East Room of the White House. It was the end of a long journey for these two scientists as well as a public show of unity after a hard-fought battle to stake claim on the first complete sequence of the human genome—the 2.85 billion units that make up our genes. Francis Collins, a physician and a devout Christian, had led the publicly funded Human Genome Project. Craig Venter, taking his cues from Silicon Valley and the tech boom of the 1990s, had formed a private company to claim the same prize. Their rivalry would accelerate the pace of work to such an extent that the date of completion arrived a year earlier than expected. It was a great day to be a scientist, and I remember watching the event broadcast over the Internet from my laboratory in Oxford, hanging on every word.

What the announcement that day meant to science paled in comparison to what it will eventually mean to the public at large. That is why the U.S. president, arguably the most powerful man in the world at that time, was making the announcement rather than a spokesperson from the National Institutes of Health or the Department of Energy, which had funded much of the 13-year project. As Clinton said, it was the completion of "the most important, most wondrous map ever produced by humankind."

This moment in many ways marked the beginning of the genomic era. Today the language of genetics has entered the zeitgeist of the modern age. References to DNA are used to sell everything from cars to computers. Genetics has become a genie of sorts, promising to grant our wishes with the magic spell of its hidden secrets. During the press conference at the White House, Clinton joked about living to 150. With advances in genetics this may actually be possible by the end of the 21st century, as our understanding of human diseases and aging expands. Almost daily, geneticists take incredible leaps forward in our understanding of ourselves.

LOOKING BACK

While much of the world that day was peering into the future, my colleagues and I were thinking about how this amazing new technology could be used to explore the past. Assembled in our small laboratory in Oxford were DNA samples from all over the world, a library of genetic information that we were using to deduce details about human history over the past 50,000 years. Our work, like that of most scientists, rested on the earlier achievements of brilliant researchers from around the world. What we had going for us was the benefit not only of their results but of the incredible new machines and techniques that had been developed for the Human Genome Project (HGP).

The HGP had begun slowly as an effort to map the large-scale structure of the genome. Beginning with a landmark meeting in Santa Fe, New Mexico, in 1986, the pace of work really gathered momentum in 1987 and 1988 when a 15-year plan was developed to undertake the sequencing. A large part of the early effort was focused on developing the technologies that were needed to plow through the huge amount of information. The HGP was gearing up to undertake something that many of us had only dreamed about a few years before, and genetics was on course to become "big science" in the way that physics—with its particle accelerators and large, international scientific consortia—had done decades before.

The pace of work was measured in the first few years of the project, as new techniques were evaluated and scientific methodologies were debated in scores of meetings and conferences. By the late 1990s, though, most of the major technical hurdles had been overcome, and spurred on in part by Venter's privately funded efforts, the HGP was cranking out huge amounts of raw DNA sequence every day. The HGP had become a massive DNA sequencing factory.

Technology was clearly no longer the limiting factor in genetic research. Rather, it was access to the genetic "texts" we were trying to read. For instance, in our global effort to piece together genetic history, DNA samples donated by people interested in their own ancestry were relatively easy to obtain from places that had a well-developed scientific infrastructure—Europe, the United States, and East Asia, for instance—but that left out most of the world. What our field of genetic anthropology needed was a truly global sampling of humanity's diversity. By analyzing samples from people who have been living in the same place for a long period of time, so-called indigenous people, it is possible to infer details about the genetic patterns of their ancestors. Furthermore, by making comparisons across many regions, it is also possible to say something about the movements of their ancestors thousands of years ago. But to do this with any accuracy, it is necessary to look at many, many people from around the world, particularly those living in relatively isolated locations.

Unfortunately, we're racing against the clock. The stories carried in the DNA of indigenous people are being subsumed into the cultural melting pot. People move for three reasons: a lack of opportunities at home, the perception of better opportunities elsewhere, or forced relocation. Many of the world's indigenous people are the poorest of the poor, residing in already poor parts of the world. Their traditional ways of life are threatened, and their children often leave home to join the economic mainstream in regional capitals or the growing megacities of the developing world. Once they enter the melting pot, their DNA loses the geographic context in which the genetic patterns create a clear trail.

The world is currently experiencing a cultural mass extinction similar to the biodiversity crisis. One symptom is the loss of languages. Linguists believe that as many as 15,000 languages may have been spoken in the year 1500, at the start of the European "age of exploration." Today only 6,000 spoken languages are left, and perhaps as many as 90 percent of these will be lost by the end of this century. We are losing a language every two weeks through the same migration process that is mixing the world's genetic lineages. While we hope that this will lead to a new sense of interconnectedness among the world's peoples, it also means that the genetic trails we follow will become hopelessly intertwined. When this happens we will no longer be able to read the historical document encoded in our DNA.

It was with this sense of urgency that we launched the Genographic Project in April 2005. A five-year, 40-million-dollar research effort, it seeks to capture a genetic snapshot of our species at this point in time, before the genetic trails can no longer be followed. It is an ambitious, international endeavor that promises to fill in the details of where we all came from. It is an attempt to answer, using the tools of genetics in concert with those of other historical fields such as archaeology, linguistics, and paleoanthropology, that key human question: Where do we come from? We hope that by the end of the project we will have a much richer answer to this question.

ORIGINS OF THE PROJECT

In August of 2002 I had recently finished a film and a book, *The Journey of Man,* about recent advances in Y-chromosome analysis. The Y chromosome, as you'll learn in this book, is a terrifically useful tool for anthropologists interested in looking at human migration patterns, and the story we told in the film was an exciting glimpse into a new scientific field.

I was traveling to promote the project and was waiting in London Heathrow Airport's Terminal 4. Sitting with me that day, across a table in a restaurant, was Kim McKay, the person to whom this book is dedicated. At the time Kim worked in senior management for the international division of the National Geographic Channel, and it was her responsibility to make sure that as many people as possible, from as many places as possible, saw the film. But her interest went deeper, and she truly thought the scientific work was amazing. Then she asked me a fateful question: What do you want to do next?

My mind raced. There were so many things, ranging from conducting focused studies of particular parts of the world to making more films and writing more books. She said that National Geographic was very interested in the work we were doing as genetic anthropologists, and she asked for my "blue sky" sketch. If anything were possible, what would the next step be?

After thinking about it for some time, I replied, "We need more samples. A lot more. What we know about human migratory patterns is based on a few thousand people who have been studied for a handful of genetic markers. There might be as many as 10,000 people whose DNA has been studied if you add up all of the samples in every paper that has been published in the past few years. But this isn't a great sample of the world's 6.5 billion people. It's like attempting to describe the complexity of outer space with a pair of binoculars. We need to increase this number by at least an order of magnitude, to 100,000 or more, to have the power to answer some of the key questions about our past. That will give us the genetic telescope we need to detect subtle migratory events in human history—and these are often the interesting ones."

We both went away from the meeting that day with the seed of something incredibly exciting germinating inside our heads. Over the next few months, in discussions with the National Geographic Society, we drew up plans for this exciting scientific venture. This would be the first time this work had been carried out using the same technologies, in the same timeframes, using the same ethical methodologies. It was a chance to do the science right. It would be open to as many people as possible, even to people whose mixed background made their genetic patterns difficult to interpret, so we would include a "public participation" component. It would raise awareness about indigenous cultures around the world and give something back to these people whose way of life was threatened. And it would allow us to share the amazing stories that came out of the scientific research in compelling ways.

At its core, though, the Genographic Project would be committed to scientific exploration—to discovering exciting, novel things about our shared past as a species. Using genetics, it is building on earlier National Geographic–funded work by such scientific luminaries as the Leakey family, Jane Goodall, and others. It is a project with many interwoven components, but at its core is the science. Without a solid grounding in basic research, the project will not have a major impact on our understanding of where we all came from. To help us on this front, IBM became involved, and their computational biology team will be instrumental in helping us to analyze the complex dataset that takes into account genetic data, linguistic patterns, the archaeological record, and stories told by the participants who have given us samples.

After much detailed planning and discussion, the Project launched in 2005 and has begun the process of collecting and analyzing DNA samples from indigenous and traditional people. The general public around the world has also been invited to participate in the study by purchasing a

Genographic Project Public Participation Kit. By sending in a simple cheek swab sample, a participant can learn about his or her own place within the story of human migration while contributing to and participating in the overall Project. As this book goes to print, almost 160,000 people have joined the Genographic Project by purchasing a kit.

This book provides an overview of what we know now: How we read DNA as a historical document, what we mean by "deep ancestry," and what we hope to learn as the project progresses. It's a whirlwind tour of a field that has developed over the past 50 years and is now poised to answer many questions that were only open to speculation in the past. We start with a brief history of genetic anthropology, then get into the meat of how to infer migratory details from DNA with stories focused around key people and their personal histories, moving back to deeper and deeper roots as the book progresses until we reach the common ancestors of everyone alive today.

I

THE BLOCK

Imagine yourself in outer space, somewhere near the moon. The Earth appears to be a blue orb floating in darkness. There are no other planets nearby—it is alone in the darkness. You begin to zoom toward it, and the lush green of the land becomes apparent. Gradually you start to make out continents—Asia, Africa, the Americas. You focus on North America, precariously connected by a narrow land bridge to South America. Zooming closer, you narrow your destination to the eastern seaboard of the United States, then closer still, rushing toward New York City. Its web of streets, railways, and bridges comes into focus, and you can make out the five boroughs that are home to more than eight million people.

New York is the world's most cosmopolitan city—perhaps the most diverse group of humans ever to live in a single geographic location in history. People from more than

180 nations (and there are only 192 in the world) have made their way to its web of crowded streets to build a new life. More languages (138) are spoken in Queens, one of its boroughs, than in most countries. It truly is America's—and the world's—melting pot.

When I visit New York, the diversity of people here is always one of the things that strikes me, along with the honking cars, tall buildings, and smell of roasting kosher hot dogs. Imagine a single city block, the diversity of stories. In one building, a Puerto Rican mother of two, struggles to raise her children and finish school; in another a Chinese immigrant family has a son on the honor roll; in an Irish Catholic family of six, the father and two sons serve on the police force; on the street, an Ethiopian cab driver is saving to bring his family over from Addis Ababa; an Italian shop owner and his wife sell deli specialties from the old country; one of their customers is a 20-something adoptee with no idea where his ancestors came from. These are typical stories from a huge breadth of geographic backgrounds, many lured by the Statue of Liberty's invocation to "Give me your tired, your poor, your huddled masses yearning to breathe free." Some say the thing that makes America great is this multiracial "melting pot" whose disparate elements combined to produce one of the most creative countries in the world.

Many New Yorkers, as in most of America, describe themselves with hyphenated words. Irish-American, African-American, German-American: People feel the pull of an ancestral homeland that most have never visited, and many know little about. This pull is strong enough to have prevented the completion of the melting process. While proud of being Americans, these New Yorkers still long to identify with something that lies beyond the Mets and the Yankees, beyond the bridges and tunnels and the estates of the Hamptons—something that binds them together in a way that the venerable constitution of a 200-year-old country cannot: a blood relationship. *Roots.*

The largest mass migration in human history took place between 1840 and 1920 when nearly 40 million people (more than double the U.S. population in 1840) moved from Europe to the United States. These immigrants included 4.5 million Irish spurred on by the devastating effects of the potato famine, 5 million Italians escaping poverty, and 2 million Jews fleeing the pogroms of eastern Europe. Nearly half of all Americans alive today have ancestors that passed through the main immigration facility on Ellis Island in New York Harbor.

Most Americans are deeply curious about their roots dating back to before their ancestors arrived in America. Like other nations with a large proportion of immigrants—Australia and South Africa, for instance—they yearn for a past left behind in the ancestral villages and towns of the old countries. Genealogy is the second most popular American hobby after gardening (and the second most visited category of Web sites after pornography). In some ways, genealogy is a national obsession.

My own interest in genealogy goes back further than the past few generations. I consider myself to be a historian, and like most historians, I am fascinated by those who lived before me. Since childhood I have been obsessed with the idea of traveling back in time, sparked at first by visits to the King Tutankhamun exhibit that toured the U.S. when I was eight, then by the history section of the local library. I spent many afternoons poring over volumes on the Greek and Roman worlds, Egyptian mythology, and the European Middle Ages. Only later did I become interested in science, when my mother returned to graduate school to pursue a Ph.D. in biology. Spending time in the lab with her, I discovered that science was really about solving puzzles and investigating mysteries, both terribly exciting pursuits for a ten-year-old boy. By the time I applied to college, I had decided to pursue biology as a career—but focusing on its historical side, genetics.

Genetics is the study of inheritance, and the field was going through a revolution when I was an undergraduate. Advances in molecular biology were making it possible to decipher the very kernel of inheritance—the molecule DNA—and to study the mechanism of why we are made as we are. Moreover, DNA also carries the story of our ancestors written in its simple code. We inherit our DNA from our parents, and they got it from their parents, and so on—right back to the beginning of life on Earth.

I decided that I wanted to devote my life to studying the historical information in our DNA, which meant focusing on evolutionary genetics, the study of how DNA sequences change over time and why. Essentially, I saw this as the late 20th-century equivalent of studying ancient languages in order to decipher inscriptions uncovered at an archaeological site. With the right samples of DNA and knowledge of how to read the text they carried, it would be possible to study the history of life on Earth.

My focus was not on genealogy as most people mean it—the construction of a family tree to show how a surname spread through North America or to confirm a relationship with a distant relative in "the old country." I was more concerned with deeper questions that would show how humans are related to monkeys, or fruit flies to house-flies—or how we are all related to bacteria. Could genetic patterns reveal details about how different species had evolved and adapted over millions of years? Was it possible to use genetic data to tell us about the origin of our own species? Could genetics possibly even refine what we know about relatively recent events, such as the genetic impact of the great historical empires?

During graduate school I focused mostly on the long-term evolution of genes, trying to find evidence of natural selection and adaptation in other species. What became clear as my colleagues and I gathered more data, though, was the importance of demography—the way populations had grown, shrunk, or moved around in the past—in shaping genetic patterns. Finding evidence for selection was tough (at least in those early days of the field), but DNA clearly had a great deal to say about the history of populations.

In a sense, I met the genealogists halfway. People constructing family trees are typically investigating events from the past few centuries, while population genetics starts there and pushes further into the past. Most of us have a sense of our family history, but eventually we all hit a brick wall. Our DNA breaks through that wall, providing a unifying path that leads from the present into the realm of deep ancestry.

TOOLS OF THE TRADE

What is DNA, this molecule that allows us to travel so far back into the past—this history book we carry around like a gift from a long line of ancestors? It turns out to be a very long, somewhat repetitive linear molecule—a tiny piece of tickertape, a bit like a continuous sequence of Morse code but with four building blocks instead of just dots and dashes. Contained in nearly every cell in your body, it is your blueprint that in theory could be used to create a physically identical copy of you. So much information is contained in this document that if the DNA inside one of your cells were stretched end to end, it would be nearly six feet long. If you took all of the DNA out of every cell in your body, it could theoretically stretch to the moon and back thousands of times—like a molecular *War and Peace*, but a thousand times longer.

DNA stands for **D**eoxyribo**N**ucleic **A**cid. It is composed of a sugar backbone—somewhat like table sugar—that binds together nucleic acid bases, another of its components (Figure 1). These nucleic acid bases, or nucleotides, are the four building blocks of the DNA molecule, and they carry its code written in their sequence along the DNA chain. The four nucleotides are known as Adenine, Cytosine, Guanine, and Thymine. Their names aren't terribly important, so we'll just shorten them to letters—A, C, G, and T. What is important is the order of these nucleotide bases in particular regions of your genome. They can help to determine your skin color, your likelihood of getting diabetes or becoming an alcoholic, your height, and all of the other physical features that distinguish you. These genes scattered around the genome tend to be between 5,000 and 50,000 nucleotide bases in length. Altogether, the human genome contains around 30,000 genes.

Most of the genome, though, is a vast expanse of DNA sequence with no known function. Some of this probably

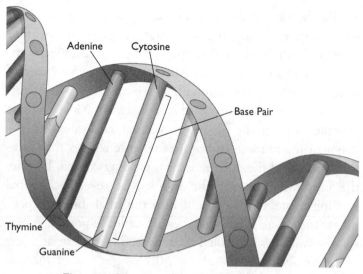

Figure 1. *Structure of the DNA double-helix.*

does have a function—perhaps in fine-tuning the way genes are turned on and off in different tissues (to make a kidney act differently from a lung, for instance)—but much probably is functionally useless. It is what geneticists tend to call "junk DNA," but it is anything but junk to those of us who use the genome as a historical document. This is our text, and it provides us with the story of our ancestors.

How does DNA work? It's actually pretty simple. If you have children, you copy your DNA and pass it on to them—a genetic hand-off that is repeated in every generation. This is why your offspring tend to resemble you more closely than they do other people. During this copying process, which geneticists call *replication,* the entire DNA molecule serves as a template to create another version of itself. There are millions of tiny little copying enzymes that do this work—think of them as monks in a medieval monastery, each copying a separate page of the world's longest illuminated manuscript—and the pieces are assembled into a book at the end. Although they are very careful in their work, they occasionally make mistakes—substituting a C for a G in one word, for instance. Most of these mistakes are caught by the molecular equivalent of the abbot of the monastery, another set of enzymes that carefully proofread the document as it is being copied. Despite the best intentions of the monks, however, occasional spelling mistakes do make it into the final, bound document. In a book these mistakes are called typos. In genetics, the mistakes are called *mutations*. They occur at a low rate every generation—about 50 changes out of the billions of nucleotides that make up the human genome. These mutations provide evolution with its basic building block: variation.

When we look at people around the world what strikes us is the incredible variability. No two people (barring identical twins) look alike. Human beings come in a huge variety of shapes, sizes, and colors. It's pretty incredible, if you think about it: We all belong to the same species, yet we look so different. Almost all of these differences in appearance originated as mutations at some point in the past, and they have been carried down to the people we meet every day.

The incredible diversity of humans led many early biologists to classify humans into separate categories. Carl von Linné—later Latinized to Linnaeus—was an 18th-century Swedish botanist who devised a classification scheme for all of the species in the world. This is known as the binomial system of nomenclature—the *Genus species* system. In the process of naming more than 12,000 other species, he chose *Homo sapiens* (which means "wise man") for us. But Linnaeus, careful scientist that he was, also went further. When he looked at people from around the world, they seemed to fall into distinct categories on the basis of their appearance. Linnaeus defined five subspecies: *afer,* or Africans; *americanus,* or American; *asiaticus,* or Asian; *europaeus,* or European; and *monstrosus,* a blatantly racist category that basically included all of the people he didn't like, including some that turned out to be fictitious. For instance, the flatheads, troglodytes, and dwarves he writes about have never been found.

These subspecies are very similar to categories that were used in scientific circles as recently as 20 years ago. In the mid-1960s, for instance, Carleton Coon—an influential American physical anthropologist—published a book called *The Origin of Races* that became standard reading for most students of anthropology. In it Coon recognized essentially the same classification Linnaeus had devised more than 200 years before, with Caucasoids (the equivalent of Linnaeus's *europaeus*), Negroids (Linnaeus's *afer*), and Mongoloids (a combination of Linnaeus's *asiaticus* and *americanus*), as well as two additional categories: Capoid (the Khoisan peoples of the southern Cape of Africa) and Australoid (the aboriginal Australians and New Guineans). What had changed in the intervening centuries was the explanation for these groups' existence. Linnaeus, taking the biblical account literally, believed that his subspecies had been created as an act of God. Coon, like all biologists after Darwin, used an evolutionary explanation: The human races had once been united, but over time they had evolved separately to produce the diversity we see today.

While Coon left out the *monstrosus* category, he did include his own racist ideology in a discussion of how Africans were trapped in an evolutionary time warp and had not yet reached the heights attained by other races. Leaving this aside, Coon's explanation for the diversity of humans was that the races had been separate for more than a million years, since at least the time of *Homo erectus*, and that traits had evolved—like those of other species—in slow, imperceptible changes that eventually produced the variety of human appearances seen today.

What was Coon's conclusion based on, though? Very little, it turned out. Anthropologists of his era were largely limited to a method used since the time of the Greeks—*morphology*, or appearance. Although morphologists measured the physical traits they studied very carefully, derived complex formulae to describe their measurements, and inferred processes from the data, they were working at a disadvantage. This is because morphological variation is ultimately produced by genetic variation, and the under-lying genetic changes required to produce a change in morphology were (for the most part) still unknown. Coon was saying, on the basis of no solid data, that in his opinion it would have taken a million years of evolution to create the differences we see in different races. Essentially, he was inferring a genetic process— evolution is, at its most basic level, a change in the genetic composition of a species over time—in the absence of any knowledge of the genetic details of what he was studying. What was needed was a way to test the genetic implications of what Linnaeus and Coon said: Did our genes suggest that humans really do come in discrete, long-separated races?

A GENETIC BOMBSHELL

The ability to sequence DNA—to determine the order of its A's, C's, G's and T's—did not exist until recently. Molecular genetics is a relatively new science because only in the past 20 years has it become possible to study DNA directly. Before the 1980s, geneticists had to rely on indirect methods to study genetic variation. These usually focused on the proteins encoded by the DNA, such as cellular enzymes. At least protein variability was closer to the underlying source of genetic variation than something as complex as skull shape or skin color.

The first demonstration that humans exhibited variation for these subcellular proteins was in 1901, when Karl Landsteiner noticed that blood from different people often clumped together when mixed. From his research came the discovery of the ABO blood groups, which were the first such inherited biochemical differences ever to be defined. These differences are caused by variation in the proteins and other molecules on the surface of blood cells. The research community soon added to these with many other protein *polymorphisms,* as they were called since *poly-* is the root of "many." By the 1960s the field of population genetics—the study of genetic variation—was awash in a morass of data. Unfortunately, although the findings should have been useful for the study of human diversity, it was unclear just how to use them.

The advances came about because of a new way of looking at biological data. Up until this point biology had been a science where long, careful observations of relatively few events (think of the insect-collecting, bird-watching biologists of the 19th century, for instance) yielded relatively simple insights into the underlying mechanisms.

Darwin used a wide variety of anecdotal evidence to support his theory of evolution by natural selection, but didn't do any direct, statistically significant tests—these would have to wait until much later. In the 1960s, though, led largely by the new science of genetics, biology started to undergo a transformation from an anecdotal science to one that strove with ever increasing rigor to apply statistical tests to data in order to better understand the underlying cause of the observations.

One of the new school of geneticists was Richard Lewontin. Since the early 1950s, Lewontin, a statistician by training, focused his work on understanding the genetic basis of evolutionary change (also the title of his seminal 1974 book). He studied data from many organisms, but focused most of his efforts on the common fruit fly, *Drosophila*. His experimental work in the 1960s revealed new ways of detecting protein variations, and his mathematical training allowed him to analyze the growing body of data in a systematic manner.

In the early 1970s, Lewontin often traveled around the country on buses. On one such trip from Chicago (where he was then a professor) to Bloomington, Indiana, he had the idea to apply some of the statistical methods he had been using to study fruit fly data to the growing body of data on human blood groups. He would use the data to test the theories of Linnaeus and Coon: Did different human subspecies exist? If they did, then most of the protein variation found in humans should be unique to the different races—it should serve to distinguish them clearly from each other.

What Lewontin found was quite the opposite, a result that surprised him at the time. It turned out that 85

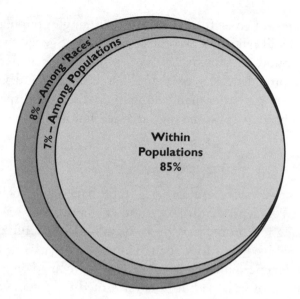

Figure 2. *Lewontin's study showed that the majority of human genetic variation is found within populations.*

percent of the variation in our species is found among individuals within a population—in other words, it is shared across all populations and races. A further 8 percent served to distinguish among populations that were members of a single race—the Dutch from the Spaniards, for instance. Only the tiniest sliver of variation—the very thin outer band of Figure 2—served to distinguish among the different races. What this meant was that human races accounted for less than 10 percent of the genetic variation in our species, and over 90 percent was found among people within a race. As Lewontin explained it, if someone were to drop an atomic bomb tomorrow, and the only group of people left alive were the English—or the Australian Aborigines, or the

Pygmies of the Ituri Forest—that single population would still retain 85 percent of the level of genetic variation found in our species as a whole. This incredible result provided clear evidence that Linnaeus and Coon were wrong. Rather than belonging to discrete subspecies, humans are part of one big extended family.

GROWING TREES

This new way of evaluating genetic data was also being applied independently by another pioneering scientist working in Europe. Luigi Luca Cavalli-Sforza, a physician who received his M.D. at the University of Pavia in Italy, never practiced medicine. His early work focused on genetic studies of bacteria but he soon turned to human genetics. Cavalli-Sforza's work in the 1950s focused on the use of human polymorphisms (such as blood groups) to understand the relationships among different human groups. While Lewontin was testing Linnaeus's and Coon's hypotheses, Cavalli-Sforza focused on using genetic data to understand more about the human past. Did our genomes contain information about how our ancestors had diverged from each other, regardless of how long it had taken?

Cavalli-Sforza began working with Oxford University geneticist Anthony Edwards in the late 1950s. The two scientists gathered published data on blood groups from populations around the world—the utility of blood groups in performing blood transfusions meant that there was a lot of data available—until they had information about the frequencies of A, B, O, and dozens of other polymorphisms in 15 populations from Africa, Europe, Asia, and the Americas. In their analysis of this large dataset, the two

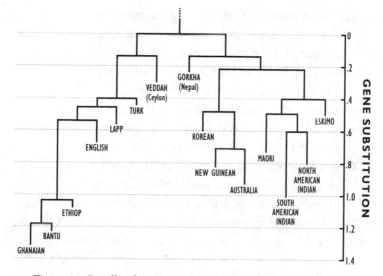

Figure 3. *Cavalli-Sforza's tree, based on classical polymorphisms, showed the relationships between several human populations.*

made a couple of assumptions. The most important was that two populations with similar frequencies of a particular blood group should be more closely related than those with very different frequencies. The second was that genetic changes tend to be rare (as we've seen), and therefore the goal in drawing a tree is to minimize the number of genetic changes required to explain the relationships. This assumption is known as *parsimony* (which we'll learn more about it in Chapter Four).

With one blood group this analysis would have been a reasonably easy exercise, but with the many blood groups they studied it was fiendishly complicated. For this reason, Cavalli-Sforza and Edwards programmed an early Olivetti computer to analyze the data for them, presaging the field of computational biology (as it is now called) by a couple of decades. What went into the analysis was genetic data, and what came out was a family tree of our species that made geographic sense—a sign that genetics could tell us how we were all related (see Figure 3). The Africans grouped together, as did the Europeans and Asians. It was a fairly basic study, compared to modern methods, but it did show that population genetics had a role to play in the field of anthropology.

One of the most tantalizing results to come out of the studies by Cavalli-Sforza and Lewontin was the finding that humans might be more closely related to each other than previously suspected. The tree derived by Cavalli-Sforza and Edwards, as with all family trees, traced back to a common source—that vertical line up at the top from which all of the other lines descend. And Lewontin showed that the genetic variation among human races was less than would have been expected if they had been diverging from each other for millions of years. The take-home message was that humans were all family.

What does this mean for everyone's concept of race? Of course humans come in a wide variety of colors, shapes, and sizes. Over the years these differences have been used to divide humanity. What the genetic data was saying, though, was that underneath the surface we are all much more closely related than we ever suspected. But exactly how close are our relationships?

From the protein data at hand in the 1970s, it was impossible to say exactly when modern humans last shared a common ancestor, and therefore how long everyone has been diverging from each other. These insights would have to wait for technological advances that allowed us to sequence DNA, which didn't happen until the late 1970s.

Walter Gilbert at Harvard University and Fred Sanger at Cambridge independently described two methods for rapidly sequencing DNA in 1977. These methods were such a huge step forward that the two scientists won the Nobel Prize in chemistry in 1980. Sanger's method became the most widely applied, since it was easier to use in the laboratory. This advance would set off the revolution in genetics that has led to the sequencing of the human genome, the biotechnology revolution, and—more germane to our story—a revolution in our knowledge of human origins and diversity.

By the 1980s, armed with DNA sequencing technology, human population geneticists were ready to start addressing the question of human diversity with greater rigor—and more data—than their scientific predecessors had ever contemplated. The work of these scientists, many of them based in the San Francisco Bay area, would provide the field with its foundation throughout the 1980s and '90s. Luca Cavalli-Sforza at Stanford University and Allan Wilson at the University of California, Berkeley, became the epicenters of a revolution in anthropology. Wilson's group would pioneer the methods for analyzing mitochondrial DNA (which we'll learn about in Chapter Three). Cavalli-Sforza's would focus on other genetic markers, particularly the Y chromosome (which we'll learn about Chapter Two). Together, their work would give us the tools and methodology to decipher the historical information in our DNA.

Likewise, the Genographic Project's goal is to assemble the most comprehensive database of human genetic variation that has ever existed. The stories contained in our DNA are told in the language of the small fraction of the genomes that differ from one person to another. By examining these regions and constructing trees on the basis of their relationships, we can infer where we came from and hopefully how we got to where we live today.

We will return to the details of the Project later in the book, but before we do we need to understand more about how we glean historical details from DNA. To do this, the story will focus on five people from around the world—people you could meet on a city block in New York, for instance. Their stories will serve as a jumping-off point for an exploration of the larger issues involved in the research. Ultimately, these combined stories will allow us to discover the roots we all carry inside ourselves—a genetic journey that will span five continents and 150,000 years.

2

ODINE'S STORY
The Exception

I visited Odine Jefferson on an early December day at his home in Fries, Virginia. This was not the Virginia of plantations and horses, but a broken-down mill town that had seen better days. The population had been steadily declining since the local textile mill closed in 1989 and today numbers only a few hundred. Most of the remaining residents have lived in Fries for generations. Odine invited me into the small house that he shared with his wife and offered me a cup of coffee. His brother, a minister in a nearby town, had come down for the day to meet me. We chatted about the odd jobs that Odine did for his neighbors, working on their lawns and house painting, and how Fries had changed over the past 20 years. Eventually we moved on to the purpose of my visit. I had come to study Odine's family tree. I asked him about his ancestors, and he told me emphatically, "I'm related to Thomas

Jefferson." Odine Jefferson, with his slight frame and toothless smile, was carrying inside him the solution to a 200-year-old mystery.

In 1998 the world was shocked to discover that Thomas Jefferson might have fathered children with Sally Hemings, one of his slaves. Although the rumor had been circulating for many years, that fall a scientific study published in the journal *Nature* appeared to have solved the mystery once and for all. The findings showed that both Jefferson's and Hemings's descendants shared the same genetic markers. Thomas Jefferson had no legitimate male heirs, so they used DNA from the male descendants of his uncle, Field Jefferson.

Geneticists then examined a part of the DNA that would have been the same in all Jefferson men. The results confirmed that the Hemings descendants had the same genetic pattern as the Jeffersons. It was as though an amateur sleuth had found the Jeffersons' family pocket watch in one of the Hemings mens' attics. While it didn't conclusively prove that Thomas himself fathered Sally's children (many suspect Thomas's brother Randolph), the genetic pattern still showed that a member of the Jefferson family and Sally Hemings had a child together all those years ago.

What was left unanswered by the original study was the elephant in the room: How had the scientists been able to pinpoint a Jefferson as the father? Why couldn't it have been another Virginia planter who had an illegitimate child with Hemings? Or, for that matter, anyone else living in Virginia at that time? The answer lay in the patterns of European DNA, our first stop on the journey back into deep ancestry.

UNTANGLING THE WEB

If you were to transport yourself back in time to late 18th-century France, when Thomas Jefferson was the United States ambassador to that country, you would find yourself in a completely different world. France was a major world power with colonies on four continents; it was the dominant economic force of the era, the most populous country in Europe, and the homeland of the world's most important language. It would be two decades before Jefferson himself, as the third president of the United States, would oversee the completion of the Louisiana Purchase from France and effectively double the size (and potential power) of his young country. France's role at that time was similar to that played by Great Britain in the 19th century and the United States in the 20th.

In France, Jefferson found himself surrounded by a whirlwind of new ideas. Many of the modern world's intellectual currents originated in the writings of the French philosophers of the time, particularly Rousseau and Voltaire. The inalienable rights of man, the social contract between governments and citizens, the triumph of rationalism over traditionalism and the divine right of kings had their roots in late 18th-century France. The ultimate outcome of these intellectual currents—the French Revolution of 1789—has been called the first true revolution of the modern era, in that it overthrew the aristocratic ruling class and ushered in a completely new social order.

Despite the forward-thinking intellectual milieu, the life of the average Frenchman at the time was very different from that of today's Parisian office worker. According to the economic historian David Landes, two centuries ago the life of an average European was closer to that of the Romans nearly two millennia before than to the lives of their grandchildren only 50 years later, and certainly to those we live today. In most ways the lifestyle of the average French man or woman had not changed substantially since the time of Caesar's conquest of Gaul in 59 B.C. People farmed the land, paid tribute (or taxes) to a local leader, and had large numbers of children, many of whom died quite young since knowledge of hygiene or medicine was limited. Their physical attributes, such as height, weight, and life expectancy, were pretty much the same as they were in the Roman era. If they traveled—which they rarely did—they did so by foot, horseback, or boat. Their lives were much more localized, played out on a far smaller field than the one we know today.

Among all of the other changes brought on by the industrial revolution of the 19th and 20th centuries, the one that had the most impact on our story is what can be called a "mobility revolution." Many of you reading this book likely had ancestors who were living on a different continent at that time. Things were not always this way, though. Demographers, who study (among other things) births, marriages, and deaths, have examined the distance between spouses' birthplaces as a way to assess changes in mobility. They have done this by painstakingly combing through church records, as any good amateur genealogist would do today. What they have found is that in the late 18th century, spouses lived only a few miles from each other, which means that most people were marrying people from the same, or perhaps a neighboring, village. Today the average distance is ten times as far; we meet and marry people from completely different parts of the world.

Figure 1. *The average distance of spouses' birthplaces has risen rapidly over the past century.*

So, people are moving around much more than they ever have. What does this mean for a geneticist studying our ancestry? The answer lies not on the right side of the chart in Figure 1—what's happened in the past hundred years or so—but rather on the chart's left side. The relative lack of mobility before the 20th century meant that people tended to stay put. Customs varied enormously from place to place, and a traveler in Jefferson's time going from Paris to the Pyrenees would have encountered a surprisingly large disparity in cultural traditions along the way. Among other things, fewer than half of the French populace at that time actually spoke French—most spoke local languages that were in some cases (Basque and Breton, for example) only distantly related to what was spoken by Louis XVI and his court. This was another reflection of the lack of mobility. Languages take a long time to develop in relative isolation—hundreds or thousands of years. The lack of a uniform language was a consequence of many, many

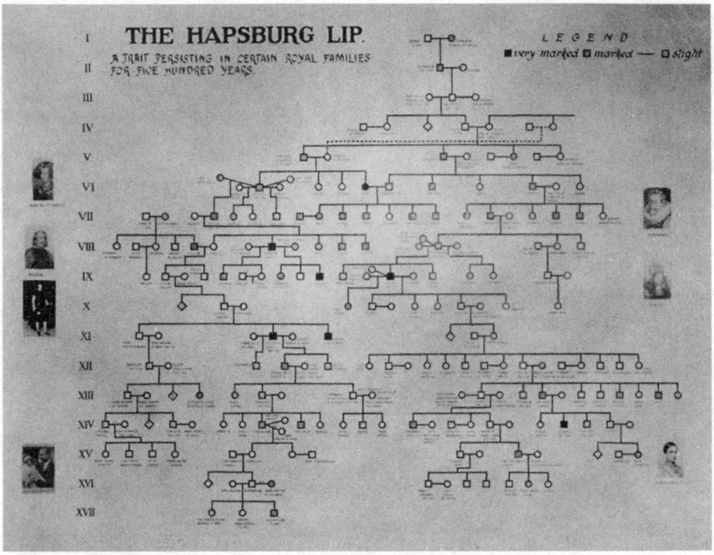

Figure 2. *Pedigree of the royal Hapsburg family.*

generations of geographic inertia. And if the people and their languages stayed put, so did their genes. Over many, many generations, our ancestors' geographic inertia meant that their genes stayed in a relatively confined location. Since we are all somewhat different genetically—unless we have an identical twin—this means that different regions had slightly different genetic patterns.

But how do individual differences from person to person add up to regional differences? Because of local marriage patterns. When the potential source of mates is limited to the few hundred people who live in your village and the one next door, you will eventually end up marrying someone to whom you are related, however distantly (first or second cousins, perhaps). When this happens your children end up sharing certain characteristics—and genes—with other people in the same region. It's as though you all had the same great-great-great-grandparents. You probably would not even realize that there is a family relationship. Yet because you share some ancestry, you would share some of the same genetic patterns.

One of the best examples of this comes not from 18th-century villages (which are difficult to study genetically without a real time machine) but from one of the royal families of Europe at that time. Among all of the great European dynasties, the Hapsburgs of Austria-Hungary were perhaps the greatest. This influential family succeeded in placing their sons and daughters into the royal households of almost every major European court, from Spain to Germany to Croatia. They even produced a short-lived emperor of Mexico. Marie Antoinette, wife of French king Louis XVI (whom Jefferson probably met as ambassador), was actually a Hapsburg.

What this Hapsburg influence meant for the gene pool is that the royal families of Europe between the 16th and 18th centuries became ever more closely related. Because they tended to intermarry as a way to consolidate political alliances, this meant that inevitably a Hapsburg would marry a Hapsburg. Over several generations, the family trees of European royalty (in effect, if a small, geographically extended, village) became quite tangled, as Figure 2 shows.

Luckily for students of genetics, the Hapsburgs bequeathed more than just their wealth and influence (and perhaps a penchant for incest) to their children. They also had a rather unusual facial feature inherited by many of them; it appeared in portraits over and over again. Marie Antoinette—despite her flattering, and apparently largely idealized, portraits—had it, as did several of her family members (see Figure 3). The Hapsburg Lip, characterized by a forward-jutting lower jaw, was a stable genetic trait caused by a mutation we learned about in Chapter One (which specific mutation is still unknown, though). The lip was pretty much unique to the Hapsburgs and became relatively common in their little "village" of interrelated family members. In fact, it was so uncommon in the general population that anyone who had it was likely to have a Hapsburg ancestor at some point in the past. It became a stable, inherited marker of that particular group—like a Scottish clan tartan or a coat of arms.

What the Hapsburg story demonstrates is that genetic markers—such as the peculiar lip—can become quite common in small groups who tend to marry amongst themselves. In genetic terms such groups are known as *endogamous*. And to a certain extent, every European village—in fact, every small geographic region in the world—was a relatively endogamous group until quite recently. Their particular patterns of genetic mutations also tended to resemble each other more closely than those from villages hundreds or thousands of miles away. Endogamy is one of the basic pieces of information we exploit to study how people are related to each other. By taking a large enough sample of people from around the world and comparing their genetic markers to those of a person with unknown ancestry, we can assign the person to a likely region of origin. We can even infer when their ancestors may have lived there.

THE RUNT

After explaining to Odine my unusual request—that I wanted to study his DNA to find out more about why Thomas

Figure 3. *A portrait of Charles II depicts the characteristic forward-jutting Hapsburg Lip, a result of his family's inbreeding.*

Jefferson's genetic pattern was unique—I pulled out a small cheek scraper, a bit like a toothbrush, and asked him if I could collect a few cells from the inside of his mouth. He said sure, and I began rubbing the scraper up and down on the inside of his cheek, first on one side, then the other.

This was the first step into the high-tech world of molecular biology. Once I had the sample safely inside the vials, it was time to get to work back in the lab. What were we so

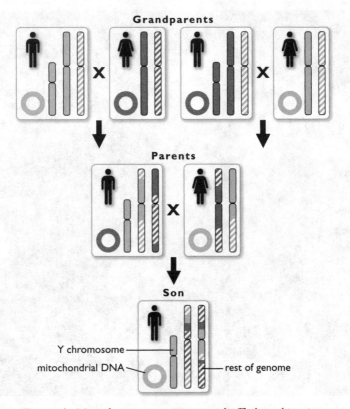

Figure 4. *Most chromosomes contain a shuffled combination of DNA passed down from both parents.*

interested in? It was a piece of Odine's DNA known as the Y chromosome. The six billion building blocks of DNA in a human are all actually divided into shorter pieces, known as chromosomes. Chromosomes vary in length from 250 million nucleotides to around 50 million, and the Y is one of the short ones. Odine was carrying a nearly identical replica of his great- . . . great-grandfather's Y chromosome, inherited down through the ages like a cherished watch, each generation passing it on virtually unchanged to the next. Occasionally there would be a scratch on the case, or a part would have to be replaced—the mutations we learned about in Chapter One—but overall it was the same watch that Thomas Jefferson had been carrying when he signed the Declaration of Independence.

Why are we so interested in Odine's Y chromosome and not another piece of his DNA? After all, the Y chromosome is just a bit player among the other chromosomal giants, with a limited role and very few functional genes. It isn't even found in women, and they seem to get along just fine without it. Why focus on this runt of the genome?

The answer has to do with a shuffled deck. All of our chromosomes come in pairs. Everyone gets one of each pair from their mother and from their father. Although each chromosome derives entirely from them, neither is identical to one of Mom or Dad's own chromosomes. That's because when a parent's chromosomes are passed on to his or her children, they reassemble themselves in a new, shuffled way (Figure 4). A little bit of chromosome A gets tagged onto a little bit of chromosome B, and vice versa along their entire length, through recombination. Why this occurs is still largely a mystery, but it almost certainly serves some evolutionary purpose or it wouldn't be so widespread in nature. Perhaps it allows the occasional bad mutation in the parent to be left out of the transmitted chromosome, or perhaps it allows new combinations of good mutations to be created. The net effect is that a child's chromosomes are entirely unique to him or her. This is a large part of the reason why no two children (apart from identical twins)—ever look exactly alike. They are carrying unique combinations of genetic variants generated through this shuffling process.

While all of this shuffling is a good thing evolutionarily speaking (to help keep the species healthy, for example), it makes the job of geneticists who want to trace ancestry very difficult. The clean lines of descent that we follow through time are quickly lost if any shuffling occurs. We do have very short regions of shuffled chromosomes (geneticists call them *haplotype blocks*) that are maintained over many generations before they are broken up, but they are very short and don't allow us to follow ancestral lines very far back into the past. What we ideally want to study is a large piece of DNA that doesn't recombine. This is precisely what makes the Y chromosome so important.

The Y is found inside cells like any other regular chromosome, but due to the vagaries of genetics, it's all alone—it doesn't have a partner like the other chromosomes. This is because it is actually the presence of the Y, mismatched with its partner the X chromosome, that determines an embryo's sex. If there is a Y present the embryo will be a boy, but if there are two X chromosomes, perfectly paired like all of the others, then it will be a girl. And because of the Y's mismatch with the X (imagine the Dickensian combination of a short, rough street urchin and a tall, genteel Victorian grande dame) there isn't any scope for recombination. The chromosomes can't believe they've been seated together at dinner and want nothing to do with each other. No recombination takes place, and the Y is passed on intact. That means the Y chromosome gives us an incredible opportunity to trace genetic lineages far back in time. We have a large amount of nonrecombining DNA that we can use to study the distribution and history of genetic lineages.

SCANNING THE TEXT

With Odine's sample in the lab, the first step was to isolate his DNA from all of the other stuff—proteins, cell membranes, salts—that might have been present in the sample we took. This is a relatively straightforward process, making use of the chemical properties of the various components of the cells from Odine's mouth. DNA is soluble in salty water, so we eliminate the things that are not by letting them settle out (sped up by using a centrifuge that spins at more than 10,000 RPM) in a salty solution. DNA also forms a crystalline structure when it is slightly dehydrated, unlike the salts with which it is in solution after the centrifuge spin. This separation is accomplished by adding 100 percent ethanol (like very strong vodka) to the solution, which then turns slightly opaque as the DNA molecules start to intertwine with each other and recoil from the low concentration of water. We then spin this mixture one more time in the centrifuge, and voilà—pure DNA on the bottom of the tube.

Once the DNA is separated, we move on to the next step: isolating the particular region of DNA that we want to examine. This can be a bit tricky and technical—in particular, how do we *know* which region to look at? The genome is a big place, composed of billions of nucleotide building blocks. If we were to sequence a person's entire genome, even devoting a full laboratory and computational power of one of the large centers that sequenced the human genome a few years ago, it would take months to complete and cost tens of millions of dollars. That's not something feasible when studying a large number of people. We need to make broad comparisons among thousands of people from dozens of locations in an effort to understand their ancestral relationships, so two or three samples won't work. Moreover, most of the information collected from each person's entire genome would be redundant anyway, since humans are so similar genetically. If you compare the same region of DNA from two completely unrelated people, you find that they are identical at 999 out of 1000 places in their nucleotide sequence. That's right—humans are 99.9 percent identical at the DNA level. This is actually far less variation (by a factor of four to ten, depending on the species) than we see in our nearest evolutionary cousins, the great apes. It reflects the fact that—like the Hapsburgs—humans are all somewhat related to each other.

The great expense of sequencing an entire genome, coupled with the relative lack of information, has led geneticists to focus on the regions where they know there is variation between people. These genetic variants, or markers, allow us to study human relationships. Because they occur so rarely and almost always originate as unique events in a single individual at some point in the past, they define unique lines of descent—like a clan. If you share a marker with someone, then you must have shared an ancestor at some point in the past.

We scan these places in the genome with known markers that will allow us to assign people to clans. In genetic terms, these clans are known as *haplogroups*—a group of people who share a set of genetic markers and therefore share an ancestor. We scan these locations in the DNA by isolating and amplifying them relative to the rest of the DNA in the genome. Typically geneticists are interested in looking at a single site in the DNA sequence, a variable position that could be (for instance) an A in some people or a T in others. We home in on the region surrounding the variable position and literally amplify it by copying it many, many times—a process known as the *polymerase chain reaction,* or PCR. It would be like choosing one sentence from the newspaper and selectively photocopying it over a billion times. By the end of the process you would have so many copies of that one sentence that the rest of the paper would become irrelevant, and your attention would be focused completely on that sentence. Detecting a change in one of the words then becomes much easier. PCR does the same thing for our DNA text, amplifying it like the sentence. It is then relatively easy, using other laboratory techniques, to determine which letter—A or T—that particular individual has at that position in his DNA "sentence."

We usually do this for a dozen or more variable positions that have been identified as belonging to particular haplogroups. On the basis of the results, we can assign an individual to a haplogroup. One haplogroup may have an A at the first site, a G at the next variable site, a C at the third, and so on. For the Y chromosome, the sites (known as markers) are typically denoted by the letter M (for marker) and a number that identifies them based on the order in which they were discovered, such as M9 or M52. The particular combination of nucleotide letters at each marker tells us to which haplogroup the person belongs. And since each changed letter is derived from an older version at the same site (e.g., for the marker M130 a C changed to a T at that particular site on the Y chromosome), we often denote them as positive for the derived state (a T in this case) or negative (a C).

Interestingly, the particular pattern of markers also tells us about the haplogroup's relationship to other haplogroups, allowing us to create a family tree of the haplogroup clans (we'll explore this more fully in Chapter Four). For now, though, we'll get back to Odine's story.

DIGGING DEEPER

Odine's DNA was analyzed for the markers that are most common in Europe, with numbers such as M45, M173, and M17, and all came back with ancestral, or negative, results. A marker known as M9 was positive, but it didn't tell us very much since M9 is found in populations from around the world—from Portugal to Melanesia to Argentina. We widened our search and eventually found that he also had a marker known as M70.

The positive M70 result, which placed him in a haplogroup clan known as K2, indicated that Odine—and therefore Thomas Jefferson—both shared a Y chromosome that is extremely unusual in Europe. At the time of the original Jefferson-Hemings study in 1998, before most of the markers we use today had been discovered, no chromosome like it had ever been seen anywhere in the world. Its uniqueness had allowed scientists to link the Jefferson and Hemings descendants in the first place. Now that more is known about global patterns of Y-chromosome diversity, we can say a bit more. M70-positive chromosomes such as Odine's are most commonly found in the Middle East and North Africa, where as many as 15 percent of the men in some populations have them. In Europe, though, they are rare. In fact, of the thousands of Y chromosomes that have now been studied in European populations, only one potential case has been reported in the scientific literature, although several have recently been found as a result of the increased DNA testing since the launch of the Genographic Project. How had one of the founding fathers of the United States, someone who looked like a completely typical northern European, ended up with this unusual genetic inheritance?

Before speculating on Jefferson's conundrum, we should ask about the other markers we typed in our quest to understand his unusual genetic pattern. How had we chosen these to focus on? We look at the most likely suspects, just like a criminal investigation. But what are the usual suspects in Europe, and why? This question—repeated around the world—is the crux of what this book, and the Genographic Project, is all about.

Our scientific goal is to explain global patterns of human diversity. To do this we need to draw together all of the existing genetic data into a coherent picture of how our ancestors populated the planet. This involves generating and compiling large quantities of data in order to recognize patterns such as the distribution of Odine's K2 lineage.

One model to follow is forensic science. The biggest DNA databases in the world are found in crime labs. The national police database in Great Britain contains nearly three million samples that have been collected and studied from criminals—the most in the world. The FBI database in the United States has more than one million samples. These databases are used to find matches in the crimes where DNA analysis can be applied. If police can collect a small sample of a suspect's skin cells, containing DNA, a match can often be found in the database. Combined with witnesses' testimony and other forensic evidence, DNA can be a powerful tool in fighting crime.

The key for any forensic DNA study is to have a good database for comparisons. Imagine trying to nab a criminal based on a comparison of an unknown sample to a database of a hundred samples, or even a thousand—the odds are pretty low that there would be a match. But as the total number of samples creeps into the hundreds of thousands, and in particular as they start to approach a large fraction of the population, then the odds of a match increase considerably.

This is similar to what we do in our studies of genetic anthropology, except that we focus on different populations. A detective using criminal databases hopes for a match; the goal is to determine the identity of a unique individual. Having a single match in the database is perfect for this sort of job, since the odds of getting one match by chance is the inverse of the number of people in the database—with a million people, it's one in a million. If the database is representative of the general population, you can even extend this to the likelihood of getting that match by chance from anyone not in the database. The key is to have a large amount of representative data—without that the exercise is futile. That's why criminal databases only started to become useful when genetic techniques became cheap and powerful enough to allow the police to generate huge amounts of data.

In the Genographic Project, we don't want to find a unique match in the database that identifies a single individual—this isn't a forensic investigation. We want to look at a representative sample of people from around the world who might be related to the person. We don't want a fingerprint, but rather a less specific measure of relationship, similar to a totem or a family crest. To make sense of what a match might mean for the person's ancestry, we need a very special database for comparisons. It needs to be representative of the entire world, but in a way that make sense historically. It's likely that we could find all of the major genetic lineages that exist in the human species in Queens, New York—but without some context the genetic data make no sense. In other words, we need a database of the way things used to be before the mobility revolution began in the 19th century—a snapshot of the world a couple of centuries ago. While time travel is not possible, we can bias toward this result by choosing very special people living today to include in our database.

Ideally they would be living in the same place as their ancestors did centuries ago. They should have been relatively isolated from immigration from surrounding groups who have moved into the region recently. They also should retain some of their ancestors' way of life, be it language, marriage patterns, or other cultural attributes. In other words, what we want are *indigenous* people (Figure 5).

Has anyone, including indigenous populations, really been that removed from the effects of global migration? In some cases, yes. I have spent time with tribes of hunter-gatherers in Africa and reindeer hunters in Siberia who—although they are very mobile and clearly don't stay in one location from day to day—have always lived in the same region. As a result, they can trace their ancestry back entirely within their group in a way that I cannot. My ancestors come from England, Scandinavia, and Holland, thus accounting for my ruddy skin and blond hair, but in no way am I indigenous to any of these places—or even to Washington, D.C., where I live.

Why is this important? Imagine in Odine's case if we had sampled someone from Egypt who had recently moved to London. Would their genes be representative of the English in a way that might allow us to study ancient patterns of migration in England, such as whether the Romans mixed with the people of the British Isles during

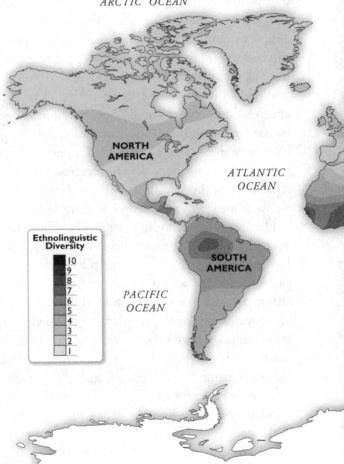

ARCTIC OCEAN

NORTH
AMERICA

ATLANTIC
OCEAN

Ethnolinguistic
Diversity

10
9
8
7
6
5
4
3
2
1

SOUTH
AMERICA

PACIFIC
OCEAN

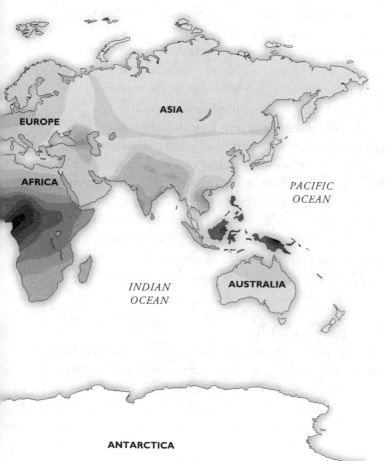

ARCTIC OCEAN

EUROPE

ASIA

AFRICA

PACIFIC
OCEAN

INDIAN
OCEAN

AUSTRALIA

ANTARCTICA

Figure 5. *A map showing today's density of indigenous populations
from around the world.*

their conquests? Of course not. Rather, it would represent another example of the mobility revolution that has become such a feature of modern life. In this example, we would exclude the Egyptian sample from our English data, since the person's deep ancestry actually lies elsewhere.

By assembling a database of genetic data from indigenous populations around the world we can reconstruct worldwide genetic patterns as they were before the mobility revolution began. While we have high-quality genetic data from perhaps 10,000 indigenous people at the moment, it is not a truly representative sample of the global population of 6.5 billion, or even the world's "indigenous population" of 350 million. Clearly we have a lot of work ahead of us, and increasing this sample size by an order of magnitude is the goal of the Genographic Project. Only with more data can we make reliable inferences about the genetic history of our species—and provide a resource to allow people around the world to learn more about their own deep ancestry.

MAPPING EUROPE

Despite the relatively limited data, we do know a fair amount about world genetic patterns. European populations are particularly well cataloged, for the simple reason that most of the world's geneticists are of European descent, and over many years they have studied themselves and the populations where they live to a greater extent than remote populations halfway around the world. While this bias makes sense, it is has led to false assumptions about the way human history has unfolded.

Clear genetic patterns exist in Europe—not all Europeans are alike. When we tested Odine's Y chromosome for a battery of markers to find out where it came from, we were employing our extensive knowledge of European genetic patterns. The seven that account for the majority of European genetic diversity—are shown in Figure 6.

The first pattern that leaps out of this mass of data is that of haplogroup R1b. Remember that a haplogroup is an ancestral clan—the descendants of (in this case) one man who had a particular set of genetic markers on his Y chromosome. The nomenclature for haplogroups is somewhat complicated, and involves a string of letters (defining a broad haplogroup affiliation, such as R) with numbers and other letters defining subgroups within the broad group (such as 1 and b) that have been assigned on the basis of the markers defining the haplogroup. Haplogroup designations such as R1b are like car models—for instance, a Volvo V70XC (analogous to the haplogroup designation) that is defined by the details of the car's components, the engine type and number, the chassis number, etc. (like the pattern of markers).

R1b is actually defined by the presence of a marker known as M343, which occurred after a long series of other genetic changes on this lineage. R1b is found at very high frequency in western Europe, where in some populations (the Irish, for instance) nearly every man is carrying it. Its frequency drops as we move eastward in Europe, and by the time we get near Poland and Hungary, it has dropped to around a third of men. On the face of it this means that the men of western Europe are, on average, more closely related to each other on their paternal side than they are to men from central and eastern Europe.

The mirror image of R1b's pattern is shown by R1a1. This haplogroup is very common in eastern Europe,

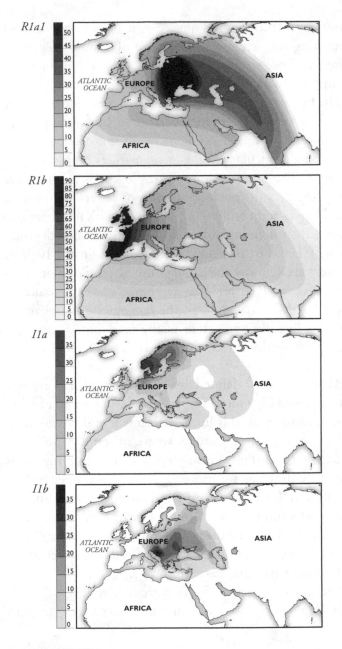

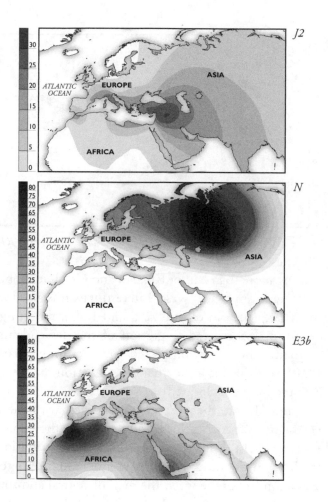

Figure 6. *Frequency distribution of the major European Y chromosome haplogroups: R1a1, R1b, I1a, I1b, J2, N, and E3b. As much as 80 percent of the European gene pool is attributed to the first Paleolithic hunter-gatherers whose descendants survived the last glacial maximum. Despite the successes of farming, only a minority of today's European lineages descend directly from early Middle Eastern agriculturalists.*

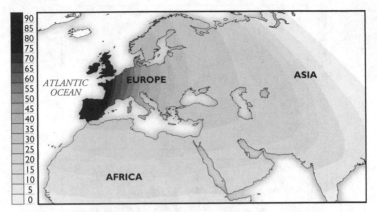

Figure 7. *Frequency distribution of Y-chromosome haplogroup R1b.*

including more than half of Russian and Czech men, but it is extremely rare farther west. Interestingly, it is also found at high frequency in central Asia and India, reflecting an ancient and widespread migration of the descendants of the man who first possessed its defining marker, SRY10831.2 (SRY is a shorthand designation for the gene where the marker is located).

I1a and I1b are the next major lineages, with the former's highest frequency in Scandinavia and the latter's in the Balkans. I1a in many ways mimics the distribution of R1b. I1b, on the other hand, is unique and attests to an ancient connection between the men of central and southeastern Europe. I1a is found at lower frequencies throughout the Atlantic fringe of Europe, while I1b is common in central and eastern Europe. These genetic connections both suggest ancient migratory connections that we'll learn about in the next chapter.

The next two patterns also are similar in many ways to each other. J2 and E3b show their highest frequencies in southeastern Europe, with much lower frequencies to the northwest. They are found at even higher frequencies in the Middle East, revealing a connection between all of these populations. The highest frequency of E3b in any European population is found in Greece, where 25 percent of the men are part of the E3b clan. E3b is found at much lower frequencies beyond the Mediterranean basin, though, and is relatively rare in many European populations.

Finally, haplogroup N occurs most frequently in northern Scandinavia and eastern Europe, and the frequency drops precipitously as we move to the southwest. This pattern demonstrates a sharp divide between Nordic and western European populations. The pattern seems surprising, but indicates a different wave of human migration.

These seven clans account for more than 95 percent of the haplogroups found in European populations. Each originated in a single man at some point in the past, and these men are truly the founding fathers of Europe. The seven clans that they spawned spread from their homelands to encompass a huge swath of territory. Who were these founding fathers, and why did their sons become so successful while others were left behind?

And what about Odine? Unfortunately, his results are still unexplained. He is a member of a Y-chromosome lineage that most likely originated in the Middle East, but the lack of comparable European samples means that it is very difficult to explain his family's pattern. It's possible that Jefferson's K2 ancestor could have arrived in Britain relatively recently (in the last few thousand years), perhaps carried by traders from the Mediterranean, but at the moment we can only speculate. We need the larger sample sizes of the Genographic Project to be able to learn more about Odine's story.

Odine aside, the data presented in Figure 7 is like the evidence unearthed at an archaeological dig. The seven genetic patterns are ancestral connections tying together men from Spain and Ireland, Scandinavia and Siberia, and the Balkans and central Europe. But it is only by arranging these genetic patterns into layers, and adding data from other fields such as archaeology, climatology, and linguistics, that we can make sense of the historical information contained in them. That is where we're headed next—to discover the methods used to order and date the layers of our genetic excavation.

3

MARGARET'S STORY
The Hearth

My grandmother Margaret's story begins like that of many Americans. She was born into a middle-class Scandinavian family near Omaha, Nebraska, in 1917. Margaret's parents had moved to America from Denmark in the first decade of the 20th century; they came, not to escape a totalitarian regime or famine, but because they felt that life in the New World offered more opportunities. Her mother, Gerda, came from Aalborg, a small town in northern Denmark. Gerda moved to America to join her sister and settled in the Danish enclave of Blair, Nebraska. Margaret's father, Emil, came from a Swedish farming family that had settled in Denmark in the 19th century. Emil was handsome—blond, with fine-boned features that contrasted with Gerda's heavier looks—and a dreamer. A cabinetmaker by trade, he was a self-taught intellectual, a socialist, and atheist who devoured much of the literature that was being written about class relations at the turn of the last century. Emil was attracted to the newness of America and spent much of 1908 as a peripatetic bohemian, riding trains around the country with hobos and taking in the possibilities.

As far as Margaret, Gerda, and Emil were concerned, they were Scandinavian—their ancestors had always lived in northern Europe. However, Margaret's story will take us on a journey far from the cold land of her parents' birth. The notion that their DNA might have linked them to another place would have been absurd, yet Margaret's DNA does tell such a story. Her legacy has been carried down to the present day—to me—in the form of a piece of genetic material known as mitochondrial DNA, or mtDNA. It is the female equivalent of the Y chromosome, and it gives us a tool to trace maternal lineages in the same way we learned about paternal lineages in Chapter Two. This tool is a bit different, though, and needs some explaining before we can get into the details of Margaret's story.

THE BUG THAT WOULDN'T GO AWAY

Biologists have been classifying life systematically since the 18th century, when Linnaeus took it upon himself to impose order on the dizzying array of life on Earth. This was the early part of the age of European colonialism, and part of the impetus for Linnaeus's gargantuan task was not just classifying species he saw in his native Sweden, but also those that were being discovered and brought back to Europe at an ever increasing rate. Naturalists needed some way of recognizing species that were related to those at

home, and Linnaeus created the modern system of nomenclature as a way of doing this. He grouped organisms
together on the basis of shared characteristics—such as the
number of fins, or the shape of the hoof—in a hierarchical
way that reflected their relationships. What Linnaeus did
not do was explain why the organisms were arranged in
this way. Darwin would show one hundred years later that
it was shared history that resulted in these groups, but
Linnaeus simply wanted to classify the plants and animals
of God's creation, and this he did, into species, genera,
phyla, and two kingdoms: Animals and Plants.

Other kingdoms, such as the Protista (single-celled
organisms such as protozoans and algae) and Fungi, have
now been split off from Linnaeus's Plant kingdom. One
group he failed completely to recognize were the Monera,
first described by Ernst Haeckel in the late 19th century,
when advances in microscopy allowed scientists to study
microorganisms in detail for the first time. Unfortunately
what Haeckel saw was probably not bacterial (it may have
been crystalline gypsum precipitated out of solution by the
addition of alcohol), but despite this the name stuck.
When microscopy further improved in the early 20th century and the incredible diversity of bacteria was revealed,
his experiments were proved to be prescient, if not scientif-
ically accurate. Recent studies using genetic techniques
have since split the Monera into two kingdoms, the
Archaea and the Eubacteria.

Scientists continue to agree on the utility of Linnaeus's
system of classification, and although the details change
as more is learned, the basic system remains the same. To
this day new plant species are described in Latin—in part
to pay homage to Linnaeus, and in part to make them

seem like the unchangeable entities Linnaeus intended them to be.

One of the underlying assumptions of the Linnaean system of nomenclature is that organisms can only be members of a single entity—it is impossible to be both an animal and a plant, for instance. Recent advances in genetics have shown, though, that many organisms—including humans—are in fact *chimeras*. We are all a combination of animals and bacteria, at least at the cellular level. And this is where we get back to the main story.

The mtDNA is actually a piece of DNA that falls outside of the nucleus where the genome proper (like the chromosomes, such as the Y) is found. The mtDNA is found inside a structure known as a mitochondrion, which lives out in the cytoplasm (or main body) of the cell. The mitochondrion has its own membrane and DNA, which is circular, unlike nuclear DNA, which is linear. This gives a clue to the origin of the mtDNA, since in nature only bacterial DNA is circular. It turns out that the mitochondrion was once a free-living bacterium that was absorbed into a cell, probably over a billion years ago, and gradually became part of the cellular machinery. The mtDNA generates energy inside the cell; to do this it uses proteins that are encoded by its DNA, as well as those found in the nuclear genome. It has a total of 37 genes doing things that are vital to the functioning of the mitochondrion, but the rest of the thousands of genes that are found in a free-living bacterium have been lost. Some of these have actually migrated to the nuclear DNA, adding to the chimeric nature of our genetic structure. In general, though, the mtDNA is a streamlined genetic package, and its 16,569 nucleotides have very little wasted space.

Since the mtDNA is found in the cytoplasm, it doesn't contain the nuclear mix of Mom and Dad's chromosomes. It traces a purely maternal line of ancestry—everyone gets it from their mothers (men have mtDNA too) but only women can pass it on to their children. This is due to the way reproduction takes place: A sperm donates only its genomic components, packed into a tight chromosomal bundle at its head. When the sperm fuses with the egg during fertilization, these are injected into the cell and find their way to the nucleus, and eventually into every cell in Junior's body. The rest of the sperm withers away, contributing nothing to the embryonic mix. Everything in the fertilized egg, including its mitochondria and other structures that perform the work necessary to keep the cell alive, come from the mother's egg.

BEDOUINS IN DENMARK?

MtDNA is a tool used to trace a line of ancestry back through the maternal line, telling us about our mother's mother's . . . mother. Let's apply this tool to Margaret: What does her mtDNA tell us about her European ancestors?

After testing Margaret's mtDNA, we found that it belongs to the group known as haplogroup J. The distribution of this group in western Eurasia is shown in Figure 1. The highest frequencies are seen in populations from the Middle East, where Bedouins in the Arabian Peninsula show frequencies as high as 25 percent, decreasing to around 10 percent in much of Europe, with slightly higher frequencies in northern Germany and Britain. Since J, as with all other haplogroup clans, originated in a single individual at some point in the

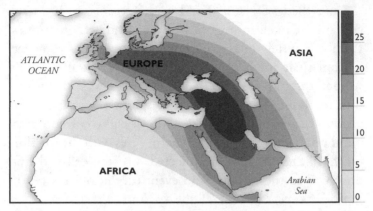

Figure 1. *Frequency distribution of mitochondrial haplogroup J*

past, this means that the offspring of this person must have spread the clan marker as she and her children moved around the world. The pattern is consistent with the migration of J clan members from the Middle East into Europe. So does this mean that my grandmother was actually a Bedouin?

Despite my love of the desert, the answer is no. The reason has to do with the date at which this movement took place, and the age of the haplogroup J lineages in Europe. It turns out that J has been in Europe for thousands of years, a result deduced from looking at the pattern of diversity within each lineage. This clock allows us to determine how old lineages are, and when they started to move from their place of origin to where they are found today.

How does the clock work? Think back to the mutations we learned about in Chapter One. These changes in the DNA sequence occur at a low but measurable rate. This is true for both the haplogroup-defining mutations we use to assign people to J, for instance, as well as at other sites on the mtDNA that change after that initial haplogroup-defining mutation took place. Imagine the defining mutation as the birth of a clan name—the very first Jones, for instance—and the other mutational changes as the given names within the clan, such as Elizabeth, Jane, Susan, and Lucy. If everyone has about the same number of children in each generation, and we count the number of names entered in the genealogical records over many generations (birth certificates or marriage licenses), then obviously the longer a clan has been around, more names will be associated with it. Clans that were founded two generations ago should have fewer members than those founded twenty generations ago, reflecting the shorter time they have had to add to the clan numbers.

A similar system of dating has been applied to the haplogroup clans we study, revealing that most are quite old—thousands, or even tens of thousand of years. In the case of J, its diversity tells us that the founder of this lineage (the very first woman to have the unique set of DNA mutations that define this clan) lived around 50,000 years ago. It also tells us that it has become widespread in Europe only within the past 10,000 years, so it's not as though a particularly successful Bedouin wandered into Denmark back in the 18th century (an event that could possibly explain Odine Jefferson's anomalous Y chromosome), leaving a genetic trail that ended up in my grandmother. Rather, the movement must have taken place long before, as the prehistoric ancestors of today's Danes were settling there. The question of how, and why, this movement occurred is where we are headed next.

ON THE TRAIL

The diversity within J tells us that it originated in the Middle East, since the members of the J clan living there have more DNA "names" than their European cousins. At some point in the past 10,000 years, members of J have expanded from their homeland in the Middle East into Europe, bringing their genetic ties to the ancestral hearth with them. The genetic pattern seems clear enough, based on the simple assumption that diversity increases over time in a predictable rate. While the pattern is somewhat interesting though, it would be nice to know why they decided to make such a move around 10,000 years ago. If they had been living in the Middle East since around 50,000 years ago, what took them so long?

The answer to this is the first example of the interdisciplinary nature of the work we do. What geneticists study is the distribution of genetic lineages such as haplogroup J and the Y-chromosome clans we met in Chapter Two. The data suggest migrations that occurred at a particular time, but why did they move from one place to another at a particular time, and how did they make the journey? The *how* and *why* questions, in addition to the *who*, *where*, and *when* that the DNA evidence provides, allow us to fill in the motives and methods of our ancestors.

To answer the how and why questions we must turn to other historical fields. Written history takes us only so far back—a few hundred, or a few thousand, years—so we must dig more deeply to find this information. The field of archaeology turns out to be our best ally in this how and why quest, since it reveals things about the way people lived, thought, and even moved. It's very much like transporting yourself back in time to pick through someone's personal belongings—what you find will reveal details about their lives, even if they aren't there to tell you about them. Imagine someone looking through your things today. Would they find a cabinet full of expensive china and an impressive art collection, suggesting that you are well-off? Or perhaps a leash and a brush full of shed hair that would reveal you to be a dog owner? Tiny clues, but in context they reveal a great deal.

Archaeologists piece together details about past lives in much the same way. The manufacturing style of a stone tool or the decorative patterns on a piece of pottery speak volumes about the people who made them. Trash dumps—known as middens—provide some of the best insights. In the same way that a private investigator may snoop around in someone's trash can to dig up evidence, archaeologists do the same, only many centuries after the trash was thrown out. In combination with other evidence, ancient trash can yield many clues about ancient civilizations.

By examining many archaeological sites in a region, it is possible to detect large-scale changes over time. While any one site may have its own unique pattern, many sites often exhibit long-range similarities due to trade or migration. It is just such a broad pattern that was discovered by archaeologists working in the Middle East in the early part of the 20th century. Their finds suggested a sea change in the way people conducted their lives—one of the most momentous turning points in human history. The famous Australian archaeologist V. Gordon Childe referred to it as the "Neolithic Revolution."

The terms Neolithic and Paleolithic are often grouped into the general term "Stone Age," because of their common suffix, "lithic" (*lithos* is Greek for stone). These descriptions were coined by the English banker John Lubbock, whose 1865 book *Pre-historic Times* described the changes in stone tools that occurred during the Paleolithic-to-Neolithic transition. Lubbock saw a significant difference between the more basic tools of the preagricultural period and those that came into use after people settled down in farming communities, as well as a general increase in the complexity of the material culture.

The details of what happened during the Neolithic Revolution have been pieced together painstakingly over the past few decades. What seems clear is that suddenly, around 10,000 years ago, people in the Middle East began to settle down and grow their own food. This happened nearly simultaneously throughout the region known as the Fertile Crescent (Figure 2), which stretches from the Mediterranean coast of Lebanon and Israel through Syria and Iraq. Several early sites, such as Çatal Huyuk, Abu Hureyra, and Jericho, have been excavated in the region, and it seems that these settlements sprang into existence in multiple locations virtually overnight.

It's difficult to overestimate the effect that this change in lifestyle had on human populations. Childe was correct in calling it a revolution, since it flew in the face of earlier theories. Since as far back as we can detect human-like creatures in the fossil record, our species had lived as hunter-gatherers. People moved from place to place, following herds of animals or rotating seasonally to the most productive gathering zones. They relied on their ability to outsmart prey and to remember the unwritten map of their territory. Because they were constantly on the move, they lived in relatively small groups—at least compared to what came next.

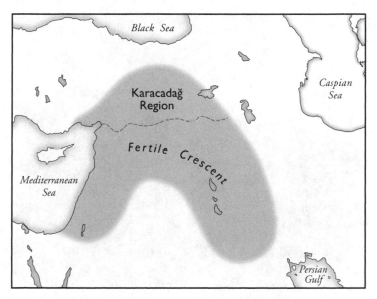

Figure 2. *Map of the Fertile Crescent. Wheat domestication likely originated in the Karacadag Mountains.*

Today, there are hardly any hunter-gatherers left in the world. Tribes such as the Hadzabe of Tanzania (Figure 3) or the San Bushmen of the Kalahari Desert provide a cultural link to the days when humans all lived in this way, but most have long since left those days behind. Not because our new way of living is better—there is good evidence to suggest the opposite, since hunter-gatherers are generally remarkably healthy if allowed to live as they always have—but because the revolution that led to agriculture allowed us to produce more people. These early agriculturalists soon became chained to their fields, trapped by their need to produce enough food to feed their growing families.

What agriculture did was to create a massive population expansion. Ten thousand years ago, at the dawn of

Figure 3. *The Hadzabe of Tanzania live in eastern Africa.*

the agricultural era, the total world population numbered only a few million hunter-gatherers spread across the world's habitable continents. Today we number more than 6.5 billion, and this is on course to increase to around 10 billion by the middle of this century. This tremendous increase got off to a rapid start. Beginning in the Fertile Crescent, agriculture soon began to spread eastward toward central Asia's river valleys, into the Indian subcontinent (although some scholars theorize a separate origin of agriculture in India), and westward toward Europe. If we look at the archaeological evidence, we see quite clearly that the spread of agriculture changed the material culture of Europe. Carbon-14 dating shows that the first farming communities in southeastern Europe date from around 7,000 years ago, while those in northwestern Europe date from only the last 5,000 years. The pattern is entirely consistent with one in which farming spread from its central origin in the Middle East into Europe over the course of several thousand years.

Two possible explanations for this spread were suggested in the 1970s. One was that agriculture, as a cultural phenomenon, was adopted by the indigenous population of Europe once they saw its advantages. At the time the very first person decided to plant seeds and settle down in one place (it was most likely a woman, since women traditionally did the gathering in hunter-gatherer groups and would have had ready access to the seeds needed to make this cultural leap), all of the surrounding groups would have been hunter-gatherers. The neighbors could have been inspired by the harvests and learned to grow crops as well, a process that could have been repeated all the way across Europe. It might have taken thousands of years for the inhabitants of the British Isles to learn about the fantastic new advance, explaining the gradient of agricultural dates we see across Europe.

The other possibility for the spread of farming, suggested by geneticist Cavalli-Sforza and archaeologist Albert Ammerman in the 1970s, is that it was the people who moved, not simply their ideas. In this scenario, the large number of descendants of the first farming communities swept aside their hunter-gatherer neighbors, spreading their own genes as they did so. If so, then we should see a genetic gradient across Europe that mimics the one for the spread of agriculture (Figure 4).

For those of you who have put two and two together already, yes—the distribution of the J clan is just such a piece of evidence that supports Cavalli-Sforza and

Ammerman's theory. This is the reason that we believe haplogroup J spread into Europe from the Middle East—one of the women in an early agricultural community around 10,000 years ago, herself a member of haplogroup J, had many, many descendants who gradually, over the course of dozens of generations, spread her clan's lineage far from the clan's birthplace. It might have even been a member of the J clan who first planted the seeds that would fuel this population explosion.

While Margaret died in 1996, my mother and I carry her mtDNA with us—a small piece of Margaret that lives on inside us today. The idea that we are also linked back to the early farmers of the Middle East is fascinating, in part because it reveals that our ancestors played a significant role in the development of European civilization. I'm sure that if Margaret were alive to hear the story, she would be similarly fascinated to learn that her ancestry stretched back far beyond the Denmark of her birth, and that her great- . . . great-grandmother was one of the first farmers. I can even imagine her saying that at long last she had an explanation for why she was so obsessive about her gardening.

But the story is a bit more complicated. First, notice in Figure 1 the unusual distribution of haplogroup J. It is found at high frequency in the Middle East, and drops in frequency moving northwest into Europe, but then becomes more common as we reach the northern coast. It is one of the major haplogroups in Denmark and Britain. We're not exactly sure why it has this unusual distribution (this is one of the puzzles the Genographic Project will be examining in more detail), but it may have something to do with the way in which agriculture spread into the

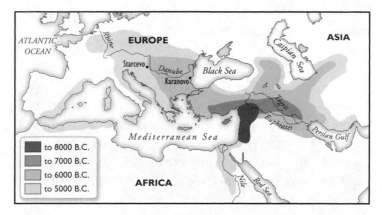

Figure 4. *This map shows the spread of agriculture during the Neolithic Revolution, showing the dates of the earliest archaeological evidence.*

northern regions. While the inland portions of Europe are quite cold and have relatively short growing seasons, the coastal areas have a more benign climate. Early agriculturalists might have lived close to the coast, where their Mediterranean crops would have been more productive. They also might have traveled by boat, which would have kept their settlements close to the coast. Whatever the reason, people like my grandmother and other Danish members of the J clan—clearly northern European in origin—have a mitochondrial link to the early agricultural populations of the Middle East.

The second factor to consider about the distribution of haplogroup J is that, while it is found in many European populations and sometimes reaches fairly high frequencies, it is still not the dominant haplogroup in any population. This suggests that the members of the J clan did not simply sweep other people aside as they migrated into Europe, but rather mixed with other clans. We need to study these other lineages to answer the question of how Europe became agricultural—the J story isn't enough.

THE REST OF THE FAMILY

Figure 5 shows the distributions of the other main haplogroup clans in Europe, the female counterparts to the Y clans in Chapter Two. In addition to J, five other clans account together for more than 95 percent of the mtDNA types in Europe. We have analyzed each of these new haplogroups in turn with the goal of assessing when they made their ancestral journeys. Comparing all of the haplogroups gives a full picture of the broad patterns of European mitochondrial ancestry.

One possibility is that all of these ancestral clans entered Europe at the same time as J, during the expansion of farming. If so we would expect all of them to show similar ages in Europe, postdating the Neolithic Revolution. By assessing their diversity we can determine if the Neolithic population expansion really did replace any hunter-gatherers living in Europe before this.

The science of assessing the place of origin and spread of genetic lineages is known as *phylogeography*. It starts from the premise that each lineage originated in a unique person—in the case of our mtDNA clans, a woman—at some point in the past. We then trace the diversity within each clan through the accumulation of mutational changes over time. As descendants of this first clan member have moved around the world, they have taken their genes with them, and we can follow the pattern of how genetic lineages relate to their geographic distribution. By assessing the relative ages of present-day clan members in different geographic regions, we can determine the direction of these ancient migrations. This is what we did for haplogroup J, showing that its pattern of dispersal is consistent with a spread over the past 10,000 years from the Middle East into Europe.

When these same dating methods are applied to the other five major European haplogroups, a somewhat surprising result emerges. Rather than dating to the last 10,000 years, most of the major mtDNA lineages are much older (Figure 6). Instead of the relatively low level of diversity we see in haplogroup J, we see far more. This is particularly true for haplogroup U, which seems to have been accumulating diversity for at least 50,000 years. Haplogroups U, H, T, and V seem to predate the arrival of the first farmers. In fact, the only lineage that appears to have arrived in Europe after 8,000 years ago is haplogroup J, as well as some less common lineages (mostly subgroups of H and T). The majority seem to date back to the time before agriculture arrived, a period known as the Paleolithic. Most European women—around 80 percent—have genetic lineages that did not spring from the Middle East with the expansion of farming. Rather, their ancestors have been living in roughly the same locale for tens of thousands of years, tracing their ancestry back to the pre-Neolithic hunting and gathering populations of early Europe. The theory of Ammerman and Cavalli-Sforza, of farming being carried into Europe on a wave of humanity, seems to have been proven wrong. Rather, it looks like most European hunters voluntarily chose to give up their existing lifestyle for a sheaf of wheat—or at least the women did. But what about the male version of the story? Can we conduct a similar analysis for the Y chromosome?

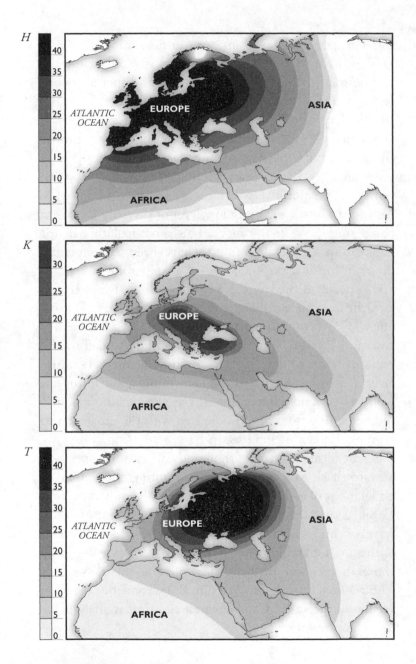

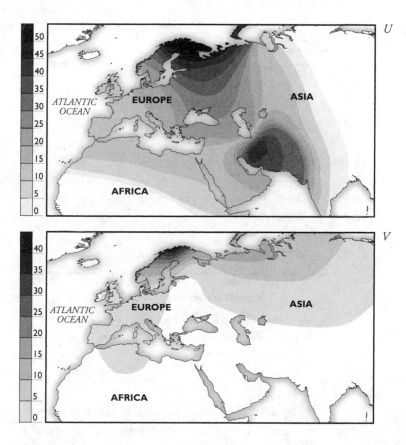

Figure 5. *Frequency distribution of five major European mitochondrial haplogroups: H, K, T, U, and V. These pre-Neolithic lineages all arose in or around the Middle East between 50,000 and 30,000 years ago before making their way into western Europe.*

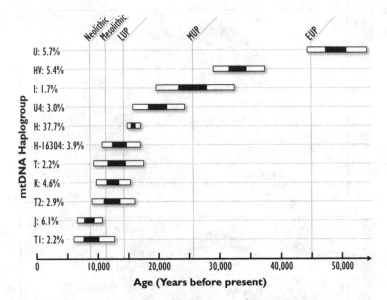

Figure 6. *Age ranges for the origin of major European mitochondrial haplogroups. White bars represent the 95 percent credible range of each group.*

STUTTERING

Most of the genetic markers that distinguish the Y-chromosome lineages in Chapter Two from each other are like those on the mtDNA—single-letter changes in the DNA sequence. These occur rarely from time to time as the DNA is copied and passed down through the generations. Because of the differences in their structure and where they are located in the cell—remember that the Y chromosome is in the nucleus while the mtDNA is out in its own little cellular compartment in the cytoplasm—the rate at which these changes occur is different as well. In general, the mtDNA has a mutation rate, as it is called, that is between ten and a hundred times as high as that of the

Y chromosome. The exact rate depends on where you are looking in the sequence, but in general it is much higher than what we see for the Y.

The rate of change for the Y is so infrequent that for many years after geneticists started to study DNA variation in humans back in the late 1980s and early 1990s, they simply couldn't find any differences in human Y chromosomes. Everyone seemed to be a genetic twin, separated at birth. Only after a concerted effort to find genetic variation on the Y chromosome did we eventually discover hundreds of locations that differ from one person to the next.

Unfortunately, while these differences allowed us to define haplogroup clans, they didn't tell us much about the level of variation within haplogroups because they are so rare. It is as though we had a set of surnames with no first names. Without being able to distinguish within haplogroups we can't apply the phylogeographic methods we used to assess the ages and migratory histories of the mtDNA clans in Europe.

Luckily, the Y chromosome does have another type of genetic variant. It is also a change in the sequence of nucleotide building blocks, but in a more subtle way.

If you are typing a word like DOG, it is pretty obvious when you make a mistake—LOG and DON are easily recognized as different words just by glancing at them. But if we type a longer, more complicated word with repeated patterns of letters, like MISSISSIPPI, it is a bit more difficult to see that MISSISSISSIPPI is a mistake without looking at it more carefully. It turns out that repeats such as this also occur in the human genome, and they cause the same sort of problem for the DNA copying machinery that long, repetitive words cause for human typists. Occasionally, when the DNA is being copied to pass on to the next generation the cellular "typist" will add or subtract a repeat. This cellular "stuttering" occurs more frequently than the single letter changes like DON, for the same reason that the first mistake happens more frequently to typists—the changes in repeat numbers are more difficult to detect.

These short, repetitive parts of the genome are known as microsatellites. They typically have around a dozen copies of the repeated nucleotides (the ISS in MISSISSIP-PI). When the DNA is copied in each generation to pass on to the next, there is a slight chance—around one in a thousand—that a repeat unit will be added or subtracted. By measuring the length of the DNA fragment using molecular sizing techniques, it is possible to infer how many repeat units are present. And because this one in a thousand mutation rate is much greater than the one in a hundred million rate for normal DNA sequences, it means that we will rapidly generate diversity in a genetic lineage. Since we know that the clan-defining change only occurred once, given the level of variation we see today we can infer how long these mutation changes have been taking place—in other words, the age of the lineage—just like we do for the mtDNA. The older a lineage, the more variation it has accumulated through this stuttering process. Lineages that are one thousand generations old will have more accumulated stuttering variation than those that have only been around for ten or one hundred generations.

When applying this stuttering analysis to the Y clans, a pattern similar to the one for the mtDNA clans emerges. Two Y haplogroups (J2 and E3b) seem to have expanded in Europe in the past 10,000 years from an origin in the Middle East. These men likely accompanied the J clan women with those first farming communities. Like the mtDNA patterns, they account for only a small fraction of the genetic lineages in Europe—also around 20 percent; 80 percent stem from other migrations. However, the Y changes reveal details that the mtDNA mutations do not.

The Y-chromosome R1a1 clan also dates from this time, but its unusual distribution suggests it spread across the steppes of southern Russia into central Europe, not from the Middle East. We don't see a similar pattern for the mtDNA, but the high frequency of R1a1 in central Europe (where as many as 40 percent of men are members of this clan) begs an explanation. The most likely answer is that the populations living on the steppes of eastern Europe were the first to domesticate the horse, and this advance may have allowed them to spread across the steppe region. It's interesting that R1a1 has such an abrupt drop in frequency as we move west into the forested regions of Europe—precisely what we would predict if the advantages conferred by horses were what allowed the clan to become so successful on the open grasslands. Consistent with this interpretation, R1a1 is also found throughout central Asia and down into northern India. It's possible that the early speakers of Indo-European languages (which include English, French, and the other languages of Europe, as well as those spoken in Iran and much of India) could have been R1a1 clan members, and their nomadic lifestyle could account for the spread of this lineage. The lack of a similar pattern on the female side suggests that conquest—a largely male-driven process since soldiers are typically men—could have contributed to the spread.

The N clan is much more recent, dating back only a few thousand years. This is consistent with its very limited distribution in Europe, where it is largely found in Scandinavia and Russia. Interestingly, as with R1a1, it has ties to regions farther east. However, the distribution and pattern of diversity of N in Asia suggests a different origin for this clan. It came out of Siberia—in particular, the region of the Altai and Sayan mountains that we will learn more about in the next chapter—and from there expanded into Europe around 3,000 years ago. Most of the people with this lineage have a similar lifestyle and language—they are reindeer herders speaking languages belonging to the Uralic family. It is likely that a burgeoning reindeer population helped them move into Europe in a same way that the horse helped R1a1's members to migrate.

The final three of the Y clans (I1a, I1b and R1b) have been in Europe much longer, dating back to around 20,000 years (in the case of I1a and I1b) and 30,000 years (R1b). R1b, then, is the equivalent of the mtDNA clan U—the oldest in Europe. It is found in the vast majority of men living in far western Europe, but is less common farther east. I1a has a similar distribution, but is somewhat younger. I1b has an unusual distribution that suggests a much different history. It is common in the Balkans but drops in frequency as we move east and west. All of these movements must have taken place before the spread of agriculture, since the patterns of microsatellite diversity predate 10,000 years ago. With R1a1 and N there are ready explanations based on animal species that might have given the clan an advantage, but this isn't the case for the R1b and I clans.

To explain these patterns, we need to examine the time when the early members of each clan lived—to try to put ourselves in the world of our Paleolithic ancestors. We have to don a different cap from the ones we have worn before as genetic archaeologists.

STORMY WEATHER

The old saying "When all else fails, talk about the weather" can sometimes be applied to scientific exchanges as well—at least those in biology and anthropology. This is because, to a great degree, the climate determines the distributions of plants and animals, including humans. Think about the incredible diversity of a tropical rain forest, with its thousands of species: colorful butterflies flying lazily through gaps in the trees, leafcutter ants carrying their green baggage back to the colony, monkeys calling in the distance, hummingbirds feeding from fragrant flowers, and the warm, moist smell of the forest itself. The wide variety exists because of a very specific set of climatological conditions: tropical warmth, large amounts of rainfall, and long-term stability. These combined factors allowed the evolution of a large number of species, each exquisitely adapted to their unique niches. An identical collection of species could never naturally exist outside of the tropics because of the unique combination of factors that are necessary for survival. There aren't many hummingbirds in the Arctic, and leafcutter ants would not survive long in the largely leaf-free environment of a desert.

Similarly, changes in climate over time have affected the distribution of animal and plant species in the same way as geography. The Earth is currently relatively warm, at least compared to long-term averages. Only 16,000 years ago we were in the most severe part of the last ice age, for instance; the climate was much different in many parts of the world. While the ice sheets that give this period its name did not stretch into the tropical region, they did cover vast parts of the Northern Hemisphere—much of present-day Britain and Scandinavia, Siberia, and North America. Only after the ice melted could animals and plants return to these regions.

The ice age provides a key to the mystery of European genetic patterns. First, looking at both the mtDNA and Y-chromosome lineages, we can detect similarities in the lineages found in Spain, western France, Britain, and southern Scandinavia—the so-called Atlantic Fringe of Europe. Similarities also exist between the Balkans and central Europe, not only for the recent Fertile Crescent lineages like mtDNA haplogroup J, but also for Y-chromosome haplogroup I1b, which is much older.

The diversity within these haplogroups tells us where they have been accumulating mutational changes for a longer period of time. Southern European populations contain more diversity for R1b, E3b, I1a, and I1b than their northern counterparts. This pattern of diversity suggests that humans withdrew to the southern parts of Europe during the worst part of the last ice age, known as the last glacial maximum. Trapped in so-called refugia, in the Iberian Peninsula, Italy, and the Balkans, they eked out a living as the ice sheets grew larger. Once the climate shifted, though, they then expanded northward to repopulate northern Europe in the wake of the retreating ice, as shown in Figure 7.

This pattern can be found in many plant and animal species, and together these two pieces of evidence provide strong support for the refugia hypothesis. The people trapped in Iberia would have had primarily R1b and I1a lineages, which then expanded into northwestern Europe where they are seen today at high frequencies. A similar pattern is seen for mtDNA haplogroup V and subsets of haplogroup H. The Balkan refugium was primarily E3b and I1b, which expanded after the last glacial maximum into central and eastern Europe.

The changing climate at the end of the last ice age seems to have had another effect on human populations. While ice sheets never extended into the Middle East during the last glacial maximum, the conditions were significantly different there from the Mediterranean climate of today. In general, this part of the world was cooler and wetter during the last ice age. The human populations living there existed largely as did other humans around the world, hunting and gathering on the relatively lush landscape. As the temperature started to climb around 10,000 years ago, however, the climate became drier. This led to an abundance of grasses, particularly those with tough outer seed coverings that could survive a long summer drought. Some human populations then began to devote much of their time to gathering these grass seeds, which could be ground to produce calorie-rich flour. These grasses were the ancestors of today's domestic wheat, gathered first in the Fertile Crescent region. Genetic studies have actually pinpointed the likely place of origin for the wild wheat strains that later were domesticated to the Karacadag Mountains in

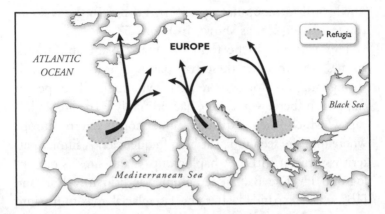

Figure 7. *During the last ice age, humans took refuge in warm parts of southern Europe. These areas served as staging grounds for the future recolonization of Europe.*

southeastern Turkey, not far from the earliest known Neolithic settlements in northern Syria.

When that first woman in the Fertile Crescent decided to plant some of the seeds she had gathered, rather than grinding them up for food, she set the subsequent revolutions in motion. Within a few thousand years of this pivotal event the human population would grow enormously as a result of the easy access to calorie-rich food, leading to the expansion of mtDNA and Y-chromosome lineages.

The ice age effects on human genetic patterns in Europe demonstrate how important the climate has been in shaping human history. From hunter-gatherers retreating from the growing ice sheets of continental Europe to early farmers growing the seeds that had grown plentiful at the end of the last ice age, climate has helped to determine human behavior and migration patterns. As we confront global warming in the present century, we should bear in mind how much our ancestors were affected by changes in temperature of just a few degrees.

TREES FROM GENES

We've learned about the spread of clans in Europe, driven by such factors as agriculture, domesticated animals, and changes in the weather. We've seen that sometimes male and female patterns don't agree, a reflection of human behavior. We've even traversed the Neolithic boundary, squarely back to the icy days of our early ancestors. Where do we go next?

We need to set our horizons higher—to encompass the entire planet. The world has more to tell us than Europe, and—as we've seen with the Neolithic connections between Europe and the Middle East—Europeans are connected by migratory history to populations around the world. DNA carries the details necessary to make sense of the dizzying array of possibilities, in the same way it helped to explain the patterns of diversity in Europe. The next part of our journey will lead back into the Paleolithic age of hunters and gatherers, into the heart of the Asian landmass, and beyond.

4
PHIL'S STORY
The Ice

I met Phil Bluehouse in 2002 when I was making the film *The Journey of Man,* in Canyon de Chelly, Arizona. His passionate and articulate explanations of the Dene, or Navajo, creation stories impressed me. They include the migration from the mother to the Earth during creation, analogous to the act of birth. Phil was particularly receptive when I spoke about migrations connecting people from different parts of the world. He told me about the pictures he had seen of people living in Central Asia and thought that he could pick out similar facial features—he even thought he had seen someone who looked like his cousin Emmett. Phil's ability to synthesize the scientific data with the traditional teachings of his people served as an inspiration as we were planning the Genographic Project.

I invited Phil to be a part of the project launch in April 2005, and he immediately accepted. He was excited about the possibility of getting his own DNA tested, since he had a hunch that he was related to people in Asia. He also hoped that his DNA's story would help us to tell the larger story of how we are all related, and even teach the world something about the Navajo themselves.

Phil swabbed his cheek with the cell scraper included in the Genographic public participation kit and sent the DNA-loaded tips off to the lab. A few weeks later the results came back, just in time for the Project's launch when they would be revealed on stage at National Geographic headquarters. After his test results were announced, Phil wept—not from fear or surprise but happiness. They revealed that his Y chromosome indentified him as a member of haplogroup Q, a lineage common in Native Americans. That meant that he did have distant relatives in Asia, as he had suspected all along, because Q is also found there. He said that he had always felt a strong connection to the people of Mongolia and Central Asia, to their way of life on the vast grasslands that, in many ways, was reminiscent of those of North America. It was as though we had united him with long-lost relatives, and his emotional response resonated with everyone in the audience. In his words, "I always knew I had family over there—now the DNA results have confirmed it."

Haplogroup Q is the major Y-chromosome clan in the indigenous populations of the Americas, including more than 90 percent of them. The microsatellite "stuttering" variation associated with haplogroup Q is similar in both Asia and the Americas, suggesting that it could have arisen in either region. Archaeological evidence suggests that the Americas were not settled until relatively recently and supports the theory that Phil's ancestors entered the

Americas from Asia. But how and when did they do it? Can DNA tell us more about these first settlers and their journey? The method for showing how Phil's haplogroup clan is connected to other lineages in Eurasia, and which populations could have settled the Americas, will be the next stop on our scientific journey.

THE BIGGER TREE

One of the most amazing sights in the western United States is the flush of fall colors in the Rocky Mountains. Groves of quaking aspen trees interspersed with pines and firs cover the mountainsides, their leaves shimmering in the breeze. As the days shorten and the nights grow cool, the trees begin to change color as the life-giving green chlorophyll retreats into the deeper tissues, leaving the other pigments predominant and turning the leaves a bright yellow color.

An aspen-covered hillside in the early fall is certainly beautiful, but if you look closely enough you will notice an odd phenomenon: the patchiness of the changes. Some contiguous stands of trees will have changed color, while other trees nearby are still green. While this could be caused by tiny variations in the microclimate—shadier, cooler sections of the topography, for instance—the true reason reveals fascinating details about aspen biology.

Aspens, it turns out, are among the largest organisms on Earth. Not each individual tree, of course—the giant sequoias of California easily top them on a trunk-by-trunk basis—but rather the entire linked organism. This massive plant can include hundreds or even thousands of connected aspen trunks that are bound together by subterranean runners. The largest documented aspen grove covers 200 acres, weighs 6,600 tons, and is estimated to be more than 10,000 years old. As the aspens mature, they send out a runner to start another trunk if they sense another section of the mountain is getting better sunlight. The aspens repeat this process again and again as they slowly creep hundreds of yards from where they originated. Despite their widespread range, all of the seemingly unrelated trunks do spring from a common source.

In much the same way, we can find connections that link apparently unrelated haplogroup clans into larger and larger "superclans." Ultimately by digging down far enough in the genetic soil, they all spring from the same source. If we go back far enough, all humans share a common ancestor at some point in the past.

This sort of genetic excavation is possible because of a leap of faith that has been tested and borne out by repeated experiments—a theory that allows us to infer historical relationships among genetic samples from the present day. The theory is like the engine in our time machine, and the DNA samples are the fuel. This theory is known as *parsimony*.

Parsimony is a way of breaking down how things happen into their constituent events until the relationships among the events are the simplest they can be. In other words, don't overcomplicate an explanation if a simpler possibility exists. For instance, it is possible that if I walk out the door of my house in Virginia to go down to the store, I might wander off track and pass through San Francisco en route. However, it's far more likely that I will head directly to the store. Likewise, if I drop the bottle of olive oil that I buy at the store, it may levitate above my head for 27 seconds before

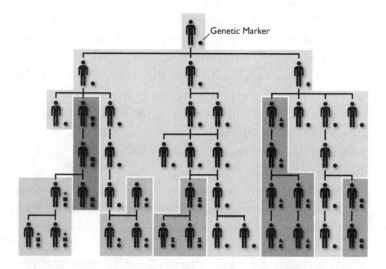

Figure 1. *As markers arise through successive generations they create new lineages that are then passed on to the next generation.*

crashing to the floor. Of course this is ridiculous—the laws of physics tell us that it is impossible—but similar laws also govern the relationships among evolving DNA sequences. If mutations occur rarely, as we learned in Chapter One, then we should assume that sequences differing at ten positions had only ten mutational changes since they last shared a common ancestor, rather than fifty or even a hundred changes (Figure 1). In genetics, as in physics, the simplest explanation is almost always correct.

Can we test the assumption of parsimony? Of course. The only way for there to have been more mutational changes than the number of observed differences is for the same position to have been changed more than once— once to a different nucleotide and then back to the original, for instance. An A became a T and then changed back to an A. In this case, we observe no changes but have actually missed two. While there are positions in our genome that do experience multiple mutations, they are easy to exclude from our analysis because they change so often that we see their mutations popping up all over the tree. They are not unlike the shuffling action of recombination we learned about in Chapter Two—evolutionary noise that makes it difficult to interpret patterns—and are a nuisance that we try to avoid.

The positions that we study to define haplogroup clans are not these hypervariable positions, but rather ones with more stable changes. We choose these positions because we want to make the assumption about parsimony, and parsimony is only valid when what you see is what you get. What we're trying to avoid are those nucleotide positions that really do pass through San Francisco in order to get down to the corner store.

So, armed with the engine of parsimony and the fuel of DNA sequences, let's try to reconstruct the history of a particular set of sequences. Imagine we sample the following sequences from three people:

Larry	A A G C T C A G G T C T A T
Sarah	A A G C T T A G G T C T A T
Juan	A A G C T C A A G T T T A T

The variable positions—the places where they differ from each other—are highlighted in grey. All of the other positions are identical. Larry and Sarah differ at only one position, while Juan differs from Larry at two positions and from Sarah at three. According to our parsimony engine, these similarities tell us that Larry and Sarah are more closely related to each other, and that the two of them are more distantly related to Juan. It's as though Larry and Sarah are brother and sister, and Juan is a cousin. If we were to draw this as a family tree, we'd get something like the following:

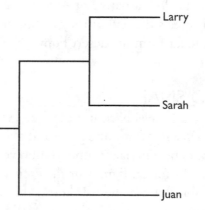

We can perform the same analysis with many, many sequences—and in fact this is exactly what happens when we analyze data from a population, or from many populations around the world. The analysis becomes very complicated with more than three sequences. With three sequences there are only three comparisons to keep track of, since there are only three possible relationships: Larry and Sarah, Sarah and Juan, and Larry and Juan. With four sequences, though, there are six possible relationships, and with five there are ten. When we examine hundreds of sequences, we have to use a computer—doing it by hand would be far too complicated.

Applying the parsimony engine to haplogroup data allows us to recognize ancient relationships among all of the world's clans. When we do this with Phil's haplogroup

Q, we find that he does indeed share a branch on the world's family tree with people living in Asia. By knowing the rate at which the changes occur, and looking at the diversity and distribution of the descendants as we did in Chapter Three, we can even say when and where that ancestor lived. Phil, a native of Arizona, turns out to have DNA that reveals the icy journey of his ancestors through a place quite far from his desert home.

A COLD SNAP

If I really could travel back in time, like H. G. Wells in his book *The Time Machine,* one place and period I would like to visit would be Siberia, around 16,000 years ago. Not for the amazing historical figures or the excitement of the culture, but for the weather. I know this sounds a little crazy—what could possibly be so exciting about such a remote land during such a remote era?

It turns out there's a method to my madness. At that time, the world was experiencing an ebb that only happens, on average, once every 150,000 years. We were at the absolute worst part of the last ice age—huge glaciers covered much of northern Eurasia and the world's temperatures were perhaps 20 degrees cooler than they are today. Woolly mammoths ruled the Asian tundra, and saber-toothed cats hunted for prey in the icy landscape. The worst temperatures during the winter may have dipped lower than 100 degrees below zero—unimaginably cold and almost uninhabitable.

Peering into the distance across the frigid horizon, it would have been possible to make out figures walking upright. Humans went about their lives in the wastelands of the far north, much as the peoples of Siberia do today.

I've lived with the Chukchi people of eastern Siberia, help-ing them as they herded reindeer, fished through small holes in the ice-covered rivers, and slept in their tents made of reindeer skin. The sheer harshness of their existence is hard to imagine. Despite being wrapped in the latest high-tech insulating layers, the cold eats into you slowly over the course of the day. By the time dinner rolls around—soon after sunset, which was at around 3 p.m. when I was there—you are ready to curl up and go to sleep, desperate to replenish the internal furnace that keeps you alive.

Why humans ever chose to live in this environment is a bit of a mystery. They may have been lured there by the mammoths and other game, or perhaps a growing popula-tion farther south made life seem crowded. Whatever the reason, about 20,000 years ago humans started to live per-manently in the coldest parts of Siberia. Archaeologists have found their tools at places with names like Dyuktai and Malta, in far northeastern Siberia, and dated them to around this time—but no earlier. It's likely that humans only learned the necessary skills to survive in the far north at this time. These would have included making warm clothing that could withstand the unbelievably cold temperatures, as well as temporary shelters that they could pack up and take with them. They were living as hunter-gatherers, since the Neolithic Revolution had not happened yet. In fact, agricul-ture has never been practiced successfully this far north in Siberia. People have always relied primarily on animals for their survival, obtaining meat and hides with which to eke out their precarious existence.

The people living in this region today, such as the Chukchi and the Yakut, probably still have lifestyles simi-lar to their ancestors thousands of years ago. They make

Figure 2. *A Chukchi family from eastern Siberia.*

use of every part of the reindeer in their lives, from the meat they eat to the hides they use for clothing and shelters, to the sinews they use to tie their wooden sleds together. While today's Siberians are herders, their ancestors of 20,000 years ago would have been hunters, although they may have practiced a rudimentary form of herding in their careful culling of certain reindeer populations.

The Chukchi (Figure 2) and other related groups have been genetically tested. Their DNA shows a link with the DNA of people living throughout Eurasia, as well as with the people living on the other side of the world in America—including Phil. Most Chukchi are members of the haplogroup Q clan.

Members of the Q lineage share the genetic marker M242—a single change from C to T on the Y chromosome—

with the people living in Siberia today. This marker originated in a single man living around 20,000 years ago, based on the microsatellite "stuttering" diversity associated with the lineage. The ancestral father of the haplogroup Q clan probably lived in southern Siberia or Central Asia, based on its distribution in world populations: About 20 percent of native Siberians have this lineage, and in some populations more than 90 percent of the men are members. The progenitor's descendants moved into Siberia soon afterward, carrying the marker for haplogroup Q with them.

As they moved to the northeast, pushing the borders of human habitation, they were entering the unimaginable cold of the Arctic Circle during the last ice age. When they reached the Bering Sea at the northeastern edge of Asia, they would have been skirting ice sheets that covered much of the tundra. Here they would have been stopped had it not been for a climatic opportunity. The growing ice sheets in the far north were locking up water that had previously filled the oceans, lowering the sea level. At the time the members of Phil's ancestral Siberian Q clan reached the region, the level had fallen so far—more than 300 feet—that the clan members would have been able to walk across a land bridge into Alaska.

When they arrived in North America, they were the first humans ever to inhabit this continent. While Africa and Asia had long lines of human ancestors going back millions of years, through *Homo erectus, Homo habilis,* and various species of *Australopithecus,* the Americas were terra incognita for people. Moreover, the vast majority of the continent was locked away behind an ice sheet that covered present-day Canada. It's likely that it was only toward the end of the ice age, when an ice-free corridor opened up along the western side of the Rocky Mountains, that the population was able to move into the prairies of North America en masse. Some early migrants may even have made a coastal journey, skirting along the great ice sheets of the Canadian Rockies and sailing down to present-day California. However they journeyed, it's clear from the genetic data that the haplogroup Q clan is no older than 20,000 years and that it originated in Asia, which means that humans entered the Americas no earlier than then. While there are disputed archaeological sites of human presence (such as Pedro Furada and Monte Verde in South America) possibly from as long as 35,000 years ago, these finds are not accepted by most archaeologists. Most agree that the earliest well-documented evidence of a human presence in the Americas dates to after 15,000 years ago.

During the entire migration from southern Siberia to the tip of South America, which probably took place in less than 5,000 years, the frequency of haplogroup Q in the migrating population also increased. By the time they reached the Bering land bridge, Q clan members probably composed close to 100 percent of the population. How this happened, by chance, is where we're headed next.

PLAYING DICE

Las Vegas casinos do a brisk business, selling themselves as family holiday destinations. The all-you-can-eat steak and lobster bars, sunny climate, and big-budget shows draw many visitors, but most people still go there for one thing: the gambling. And despite what people may think about their sure-fire "system" for winning, or the "hotness" of the slots, the casinos turn out to be the big winners. Each game is slanted ever so slightly in favor of the house. The most popular games—slot machines—have a house advantage as high as 25 percent, meaning that the house is 25 percent more likely to win than would be expected by chance alone. In blackjack the house advantage (with a perfect playing strategy on the part of the gambler) is less than 1 percent. Assuming that quite a few of the players will make mistakes, it may actually average out closer to a few percent. These advantages are not enough to deter the individuals who feel lucky (or smart) enough to play, and they continue to step up to the tables in droves. Summed over the millions of people who play every year, the small house advantage is enough to guarantee a steady stream of income for the casino owners. Players lose an average of six billion dollars a year in Las Vegas alone. Overall, the U.S. gambling industry takes in more money each year than movies, spectator sports, and recorded music—combined.

Clearly, the perception that an individual can win is important in the equation. Although gamblers may realize that overall the odds are stacked against them, they also know that any given game could go their way. This disconnect between the individual event and the sum of many events is a result of a statistical property known as the law of large numbers. If you think about flipping a coin, on average one out of every two times it should come up heads, since there is a 50 percent chance of getting this result. On any given flip you can get either a heads or tails, regardless of what happened before. The law of large numbers says that, summed over many flips, the proportion of heads and tails approaches 50 percent with absolute certainty, as though the coin "knew" what had happened before.

With a smaller number of coin flips, however, the proportion can be quite different from 50 percent. If we flip the coin ten times rather than a thousand, we may get seven heads and three tails, or four heads and six tails. This doesn't mean that the likelihood of flipping heads or tails is different from 50 percent, only that we have not sampled very many coin flips in our experiment. The error introduced by using a small number of events is known as sampling error.

What does all of this have to do with genetic lineages? The answer is that human populations usually behave like coin flips. The likelihood of passing on one's genes varies from person to person, but if we have a large enough number of people, then it tends to even out in such a way that the variation distinguishing the population is transmitted to the next generation in the same proportions seen today. The population of Europe is now quite large (the European Union alone has 457 million people), so we expect the proportions of the lineages in 2030, a generation down the road, to be similar to today.

In small populations, though, the frequencies of genetic lineages can vary quite a bit from one generation to the next. Just as in the coin-flipping example, if a small band of hunter-gatherers contains only ten people, then their children can end up with quite different proportions of lineages from their parents. Unlike the coin-flipping example, though, the new frequencies change the likelihood of passing on the lineages to the subsequent generation. It's as though we flipped seven heads and three tails, and then the likelihood of getting heads changed from 50 percent to 70 percent. This intergenerational ratchet-like behavior occurs because the lineages that are passed on to the next generations are sampled from the present generation.

You can see from this example that the frequencies of lineages in small populations can change quite a bit in a very short period of time. One lineage may even increase in frequency until every single person in the population is carrying it. The change in frequency from generation to generation due to sampling error is known as *genetic drift* (Figure 3). There is no outside force, like natural selection, that causes the frequencies to increase or decrease—the lineages drift around randomly from one generation to the next due to the whims of chance.

This phenomenon is what appears to have happened to Phil's ancestors as they were moving into the cold northeastern reaches of Siberia. The number of people who can be supported through reindeer hunting is relatively small, and the groups were probably not much larger than 25 people—about the size of the average Chukchi group today. Over time, as the populations moved farther and farther into unsettled territory, splitting from each other and forming new groups, genetic drift worked its magic on the lineage frequencies. By the time a small group reached Beringia, the frequencies had changed substantially since the first group headed north thousands of years before. We know this because so few other genetic lineages made it across into the Americas. Today's Native American populations are remarkably similar to each other genetically. Phil's lineage, Q, is found in Native American men living from Alaska to Argentina, and along with its descendant Q3, it is pretty much the only Y-chromosome lineage found in South America. In western North America there appears another lineage, C3, which was brought to the continent in a later migration that never reached South America. Together these two lineages (C3 and Q) account for 99 percent of Native American Y chromosomes. In addition, there are only five major mtDNA haplogroups (A, B, C, D, and X) in Native Americans. This is in contrast to the dozens of mtDNA and Y-chromosome lineages found in Eurasia and Africa. The discrepancy shows the effects of genetic drift on the populations that endured the worst part of the Siberian ice age to migrate into the Americas.

But what about all of the other lineages in the ancestral Asian population that gave rise to Q, C, and the other lineages—where did they come from? Can we look even further back, beyond Siberia, for Phil's more distant ancestors—in essence, to derive a family tree that connects Phil with all of the people in Asia?

GOING DEEPER

In the same way that Phil's Y chromosome shares an ancestral marker that makes him part of the Q lineage, so too do the other lineages we see around the world. They are not separate entities, but join together into a tree based on their shared markers. Think back to Larry and Sarah sharing certain markers that link them to each other before they link to Juan—the alphabet soup of haplogroups also share markers that define deeper relationships among them, uniting them into "macrohaplogroups." One of these markers unites Phil's lineage with those from the other end of Eurasia—the marker known as M45.

M45 appears to have shown up in a man living in Central Asia around 40,000 years ago. We know he was living in this part of the world because the diversity

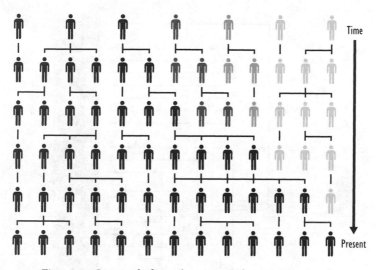

Figure 3. *Genetic drift can have a significant impact on the frequencies in a population. Here we see how three individuals (in black) could give rise to all of the present-day lineages.*

associated with M45 is highest there, and only in this region do we see representatives of all of M45's descendant lineages (Figure 4). These include P*, or individuals who have the M45 marker but no additional markers, as well as those belonging to Q and R lineages. The Q lineages appear to have originated in the same region, and link Phil into the P "superhaplogroup." Interestingly, Phil's Q lineage also turns out to be a cousin of the R haplogroup, since both lineages share a grandfather in M45. We learned about the R lineages earlier, with their high frequencies in Europe (groups R1a1 and R1b).

What this tells us is that the ancestor of most western Europeans *and* Native Americans was a man who lived in Central Asia around 40,000 years ago. From here his descendants moved westward into Europe and eastward

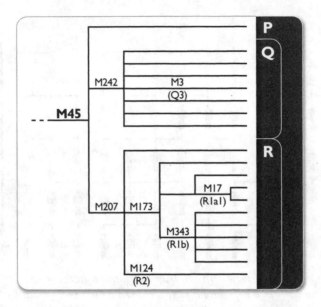

Figure 4. *The M45 branch on the Y-chromosome tree, shown here, gave rise to most western Eurasian and Native American lineages.*

until they encountered Beringia and ultimately the New World. It is poignant to consider that when Columbus encountered Native Americans on his 1492 voyage, he was reuniting two branches of the human family tree that shared the same great-, great- . . . grandfather on the steppes of Asia 40,000 years before.

DEEPER STILL

The tree of Y-chromosome clans that allowed us to unite Phil's lineage and western Europeans can also be extended to include others in Eurasia. Using the same method— looking for a marker we share in common with other lineages—we can create an even larger tree. The next such marker is one we call M9, which brings most of the clans in Asia into the tree.

As you can see in Figure 5, M9 unites a huge number of seemingly unrelated lineages. Like other haplogroup clans, they have been assigned letters to distinguish them—K through O, and then the additional lineages we already united using M45. M9 pulls together nearly all of the diverse lineages found in Eurasia, for they all share this marker.

Based on the accumulated diversity, the ancestral clan founder—the man with the first M9 mutation on his Y chromosome—likely lived around 40,000 years ago. Every one of the diverse lineages in the tree is a descendant. One of these lineages, haplogroup K, is itself a "superhaplogroup" like P. There are sublineages of K, but by and large K unites the disparate lineages of Eurasia.

Given the widespread distribution of K, it probably arose somewhere in the Middle East or Central Asia, perhaps in the region of Iran or Pakistan. From there the descendants fanned out across the continent, with N, P, Q, and R occupying the northernmost portion, while K and L stayed in the south, and M and O headed east. Today, K and L are found mostly in the Middle East and India, while M is only found in Melanesia and Oceania, and O is the predominant haplogroup in East Asia.

Like the routes of the European clans, the routes followed by these early wanderers were dictated by geography and climate. The great mountains of Central Asia, radiating out from the Pamir Knot in the center (located in present-day Tajikistan), served to split human populations as they moved through the continent. Those who headed to the north would eventually end up in Siberia, Europe, and the Americas, while those who headed south would populate South and East Asia. (The details of the routes followed by these early Paleolithic explorers in their settlement of the Eurasian landmass are discussed in greater detail in the Appendix.)

Do the patterns we see for the Y chromosome also apply to the mtDNA lineages? In other words, was our M45 clan founder of 40,000 years ago accompanied by a woman who also gave her mtDNA daughter lineages to the Americas and Europe? For the Americas, the answer is a resounding yes. The five founding Native American mtDNA lineages (A-D and X) have all been found in Central Asia, and these lineages clearly moved from there into the Americas along the same path as their male clan members.

The European situation is murkier. Although we see connections between western Asia and European mtDNA lineages, it is not at all clear that these lineages entered Europe from Central Asia. This may be due to incomplete sampling, or perhaps incomplete knowledge about the extent of mtDNA diversity in Central Asia (only a few hundred Central Asian samples have been studied, as opposed to tens of thousands of Europeans). At the moment it appears that most mtDNA diversity in Europe came from the Middle East, while the large R1a1 and R1b clans on the male side have an ultimate origin in Central Asia. Reconciling these stories is one of the goals of the Genographic Project.

What is clear, though, is that all of the present-day European genetic lineages migrated there in the past 40,000 years. However, when the earliest immigrants arrived in Europe during the last ice age they were not alone. This ancient encounter is where we're headed next.

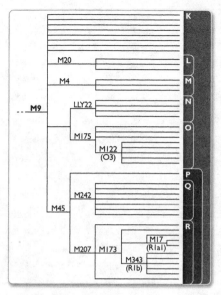

Figure 5. *The descendants of a man who bore the M9 marker and lived about 40,000 years ago gave rise to most Eurasian lineages.*

WE ARE NOT ALONE

As modern humans moved westward across the frozen wastes of the Asian steppes into Europe, they eked out a living from the animals living on the grasslands—likely hunting mammoth and reindeer and staying in nomadic encampments. By the time they reached the end of the grasslands—which during the last ice age stretched all the way into eastern France—they would have been well adapted to that lifestyle. Consummate hunters, their way of life honed by thousands of years of life in the cold north, they were capable of surviving against incredible odds. The same skills that allowed their cousins to expand into Siberia and the Americas would have prepared them for Europe as well (Figure 6). But

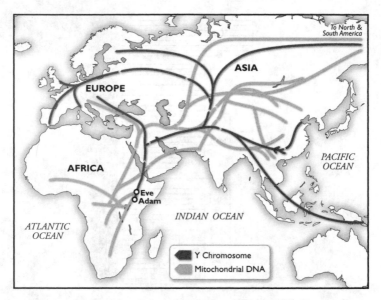

Figure 6. *As early migrants headed out of the Middle East and into central Asia, some headed west into Europe while others continued on to populate much of East Asia.*

Europe presented a different challenge, in the form of a distantly related cousin.

At the time the ancestors of modern western Europeans—the members of the R1b clan—started wandering into Europe in earnest around 35,000 years ago, they encountered another hominid living there—the Neandertals. The fossil record shows that Neandertals first appeared in Europe around 200,000 years ago, and they were well ensconced there by the time modern humans tiptoed into their territory. Life in the intense cold of the European ice age had molded Neandertals into highly cold-adapted organisms, with their thick trunks and heavy bones telling of a life of hardscrabble survival. In contrast, the incoming modern humans were relatively tall and thin, using cultural adaptations such as warm clothing to adjust to the cold northern climate.

Neandertals evolved from a human ancestor who lived around 500,000 years ago. This hominid, known variously as *Homo antecessor, Homo heidelbergensis,* or archaic *Homo sapiens,* first appeared in Africa. The descendants of this ancient human-like creature subsequently left, moving into the Middle East and Europe and steadily adapting to the colder climate they encountered over tens of thousands of years.

When the relatively lithe modern Europeans encountered their cousins on the snowy fields of central France, what was going through their minds? Did they see a kindred spirit, a long-lost cousin with a heavier-set body, or did they turn up their noses at the Neandertals' brutish ways? This brings us to one of the great debates of modern anthropology. Could the Neandertals be the ancestors of today's Europeans?

Neandertals were the very first fossil hominids ever discovered, in Germany in 1856. At the time many scientists thought that their thick browridges and other nonhuman features marked the bones as those of a diseased human, but the discovery of other similar finds, coupled with Charles Darwin's great insight into the details of evolution, eventually led to the realization that Neandertals were more likely an extinct species related to humans. But does that mean that they were our ancestors?

Debate over this topic has raged for many years, with paleoanthropologists arguing vehemently for or against the possibility. Fossil finds such as those from Altamira in northern Spain suggest that there was some interbreeding between the newcomers and their Neandertal neighbors, with skeletons showing characteristics of both groups. However, the general limitation of paleoanthropology— the small size of the fossil dataset, and any unique find, may not be representative of the general pattern at the time—meant that these discoveries were highly controversial. What we would ideally like to do would be to test the DNA of Neandertals and modern Europeans to see if the latter could have descended from the former—the sort of blood relationship that only genetics can decipher.

The problem, of course, is that Neandertals no longer exist (despite what you may think of that annoying person in the office down the hall from you). They went extinct around 30,000 years ago, so our only knowledge of them comes from the fossil record. Since our studies of human genetic ancestry are limited to individuals with intact DNA, typically this means that we can only study living creatures. Yet in some cases, we can actually extract intact DNA from long-dead tissues and obtain enough data to allow us to study their genetic relationships. In 1997, such a study was carried out on the very first Neandertal find—the one discovered in the Neander Valley in 1856—by geneticists Matthias Krings and Svante Paabo. Their paper caused a sensation in the anthropological community and helped to solve the enduring mystery of the Neandertals.

Krings, then a graduate student in Paabo's laboratory in Munich, painstakingly extracted DNA from a small arm bone specimen. They spent more than a year studying it. Early experiments were foiled by contaminating DNA from modern humans, such as people in the laboratory or the remnants of other experiments, but eventually Krings managed to piece together enough mtDNA to make a statistically significant comparison to the modern human database. From this sample, he discovered that the Neandertal sequence fell well outside the range of variation seen in modern humans—it was so different that it must have been accumulating mutational changes for longer than modern humans had. His calculation yielded a date of around 500,000 years for the date of divergence for the Neandertal and human genetic lineages. Clearly, modern humans were not the descendants of Neandertals, but rather their distant cousins (Figure 7).

One criticism of the research was that not enough mtDNA sequences had yet been obtained to validate the findings. What if the frequency of Neandertal lineages in the modern European gene pool was so low that geneticists simply hadn't found them yet? This dispute has now been settled by one of the early "discoveries" of the Genographic Project. Of the more than 100,000 individuals (most of European ancestry) who have had their Y chromosomes and mtDNA tested, none have lineages that fall outside of the normal range of human variation. Rather, every sample belongs to one of the human lineages established in earlier studies. The power of such large numbers has provided a definitive answer to the question of modern European origins and is helping us to increase what we know about the distribution of Y-chromosome and mtDNA lineages in Eurasian populations. Modern Europeans descend from ancestors who entered the region within the past 40,000 years, not hundreds of thousands of years ago with the Neandertals.

What seems to have happened is that modern humans out-competed their European cousins, driving them to extinction. Humans' greater brain power, coupled with the skills learned during their long sojourn on the steppes

Figure 7. *The first humans into Europe during the Paleolithic, the Cro-Magnon (left), ultimately replaced their Neandertal predecessors.*

of Central Asia, gave them the advantage they needed to hunt game and find other food sources. The Neandertals—trapped by their biological specializations—weren't able to adapt quickly enough to the increased competition from the newcomers. This advantage in cultural adaptation seems to have allowed the incoming steppe population to beat the Neandertals on their home turf. Anthropologists have calculated that if the Neandertal death rate increased by just one percent per year, or the birth rate decreased by merely one percent, our burly cousins would have gone extinct in just a thousand years.

Despite their advantage over the Neandertals, the newcomers would be eventually forced into retreat as well by the cold European climate. Humans, like most other animals, sought out the best place to survive as the climate changed. As large ice sheets advanced across Scandinavia and Britain, the human population beat a retreat to the best possible place to sit it out—the beaches of the Mediterranean. As we learned in Chapter Three, humans ended up in Spain, Italy, and the Balkans, where the ice was kept at bay by the warm waters of the Mediterranean.

But there was a stroke of good luck. The climate, for reasons that still aren't fully understood, started to change around 15,000 years ago. The weather warmed up, and the ice sheets began to retreat. Taking advantage of the situation, human populations started to advance back into northern Europe and took their genes with them. The current distributions of the European genetic lineages are largely due to this "overwintering" and subsequent expansion, coupled with the migrations of new haplogroups importing agriculture.

ONE FAMILY, MANY FACES

As humans were being pushed into a corner in Europe, their cousins in Asia were enduring a similar onslaught. Although small groups did live in the far north, for the most part Asians were forced into their own enclaves during the last ice age. Glaciers spreading from mountain ranges—the Hindu Kush, Tien Shan, and Himalaya—

hemmed them in to different corners of the continent. Indians were isolated in the subcontinent, and East Asians were restricted to their core territory farther east, perhaps in the area of Vietnam, Cambodia, and southern China.

Because humans at this time lived in small hunter-gatherer populations, genetic drift served to change gene frequencies over time, as these small groups split off from each other and migrated into new territory. Genetic drift, for instance, explains most of the genetic patterns we see for the mtDNA and Y-chromosome markers we've focused on so far. Drift would have led to divergence between the populations hemmed in by the ice age and mountains, helping to explain why East Asians, Indians, and Europeans all look different from each other.

Some genetic differences, though, are probably due to the action of selection. Darwin's evolutionary force, "survival of the fittest," modifies gene frequencies over time in response to environmental changes. This is currently the best theory for why human skin comes in different shades, as we'll learn in Chapter Six. It also helps to explain some changes in body shape, with people in cold, northern climates typically having stouter bodies than those in the tropics—an adaptation that helps to reduce heat loss because the relative surface area of a stout body is lower.

Another force that Darwin discussed in his book *The Descent of Man* probably also played a critical role in creating the diversity of physical appearances seen around the world today. Sexual selection, as it's known, results from choices we make about the appearance of our mates. The peacock's tail is one of the best examples of sexual selection. This enormous, feathered feature is actually quite difficult to deal with in everyday life; males with no tails are better able to move around and escape from predators. However, when the mating season comes around the tail suddenly takes on primary importance because females will mate only with a peacock that has a luxuriant tail—tailless peacocks never get a chance to pass on their genes. Over time, the choosy females have selected for the males with enormous, unwieldy tails, and now all male peacocks have them. No one knows why females first chose males with exaggerated tails. They may have served a signal of male health and vigor—if he could survive with the handicap of a huge tail, he would probably produce healthy offspring.

While humans don't have enormous tails, many of the features that distinguish us arose as idiosyncratic, local decisions about what is attractive, made by our ancestors thousands of years ago. This theory would help to explain our enormous physical diversity despite relatively little underlying genetic divergence (remember Lewontin's result in Chapter One). The theory has never been properly tested, though, since we still know very little about the underlying genetic changes that determine human appearance. As these genetic markers are discovered in the next few years, though, we should be able to test the theory that humans diverged in their surface features due to sexual selection. Exploring this idea is one of the goals of the Genographic Project.

As humans moved through Eurasia, then, the forces of genetic drift, climatic adaptation, and sexual selection combined to change their physical appearance. At the same time, other changes were taking place—changes in the languages they spoke, as well as cultural developments that allowed them to survive in their new locations. Southeast Asians would never need to create warm, fur-lined clothes, and Siberian hunters would not need to worry about sunburn most of the year. The early wanderings of our species helped to mold humanity's original African features—from language to skin color—into local, specialized varieties, producing the wide diversity we see around the world today. Most of the differences that distinguish us today probably arose during this period, within the past 40,000 years.

While all of this was going on in Asia, though, there were humans already living in another part of the planet. In fact, its fossil record suggests that it is the oldest continuously inhabited continent outside of Africa, with evidence for a human presence starting around 50,000 years ago—long before Asian steppe hunters colonized Europe. The place is Australia, and the details of how and why it was colonized at such an early date reveal a fascinating story in humanity's journey around the world.

5

VIRUMANDI'S STORY
The Beach

I met Virumandi in a small village in southern India in 2002. His contact details had been given to me by my friend and colleague, Professor Ramasamy Pitchappan, who is now the director of the regional center for the Genographic Project in India. Professor Pitchappan and I have been friends since my Oxford days in the late 1990s, when he approached me at a social event and asked if I remembered him from an e-mail exchange a few years earlier. He expressed an interest in collaborating, and we had a long and very enjoyable discussion about ways we could work together, which eventually led me to India.

Virumandi is a member of a group known as the Kallar, from the Piramalai region near the regional capital of Madurai. Pitchappan had studied samples from Virumandi's village as part of a larger survey of genetic patterns in southern India. Virumandi's village was a good place to look for these patterns, since the Kallar people have inhabited southern India for thousands of years.

The issue of Indian genetic patterns particularly interested me, as I had spent much of the previous few years sampling and genotyping Y-chromosome markers from populations in the rest of Eurasia. This work showed that Central Asia had played a key role in the populating of the world. It served as a harsh school for our ancestors, helping them to learn the skills that they would need to survive in lands far from the tropical Eden of their birth. They wore warm clothing, developed new tools, and ultimately made the cultural adaptations necessary to occupy Europe and the Americas during the last ice age.

One nagging problem stemmed from the fact that the earliest evidence for humans and their tools outside of Africa had been found not in Asia, but in Australia. Africa clearly holds the oldest evidence for our distant ancestors, from our seven-million-year-old hominid relative *Toumai,* found in the remote sands of the Chadian desert in 2002, to the earliest members of our genus, *Homo,* in the Rift Valley.

While Africa is clearly where our ancient stock comes from, the Genographic Project wants to figure out when and how our ancestors occupied the rest of the world. Our ancestors *Homo erectus* left Africa around 1.8 million years ago, settling in the tropical and subtropical zones of Central and East Asia. They died out approximately 100,000 years ago, except perhaps on remote islands in southeast Asia. The ancestors of the Neandertals left Africa around 500,000 years ago, but again their line died out. Today, humans are the only hominid species to leave descendants down to the present day. Even the early *Homo sapiens* who made it into the Middle East around 110,000 years ago died out 30,000 years later. Here is where the fossil record throws us a curve ball. The next hominids we find living outside of Africa lived in Australia, around 50,000 years ago. The hominid presence in Australia is the oldest in the world, predating the presence of humans in Europe by at least 10,000 years. But the mystery of how people migrated there is only now beginning to be explained. As we will see later in this chapter, the DNA carried by Virumandi and other Kellars helps us to sort out this mystery.

THE WASP'S CLOCK
Most of what we know about our ancestors' way of life can be gleaned only from the artifacts they left behind. By studying the pattern of marks on a sherd of pottery, for instance, we can point to something about the regional influences on the person who made it. Tools can tell us something about the level of material culture of the person who made it, and by inference the other members of his or her group. It is a way of using stone to delve into the psychology of the person who made the artifact.

Often artifacts are the only evidence that a person has lived somewhere. Human tissues, including bone, break down over time when exposed to the elements. Only recent burials yield intact soft tissue, and even bone usually decays after thousand of years. For this reason, the most useful way to track our ancestors is to uncover their artifacts. Tools obviously have to be made by someone, and since we are the only species on Earth that creates complex ones, then whatever we find must have been made by our forebears.

Our ancestors also left behind another unique relic, which abruptly appears in the archaeological record around

50,000 years ago. It is then that art makes its first appearance—depictions of humans, animals, or geometric shapes, scratched or drawn on stone with pigments. This advance, a complete break with everything that came before, gives an insight into the development of the human brain. Its relevance in tracing human migrations is that, if you find art somewhere, then modern humans must have been living there. Our distant cousins, the Neandertals, did not create artwork such as that drawn by modern humans at Chauvet or Lascaux caves in Europe, and *Homo erectus* didn't scratch depictions of the animals they hunted onto cliff faces in East Asia. We are the only species that has ever done such a thing.

Rock art, as it is called, is found worldwide. The oldest depictions are found in Europe, probably because the caves in which they were created are the perfect environment for preservation. Rock art is also found in Africa and Asia, dating to a somewhat later era, perhaps because most of it was created on exposed stone faces.

Rock art has also been found in Australia (Figure 1), and here is where the story gets interesting. Drawings discovered at Kimberley, in northwestern Australia, have depictions of human and animal figures. They were clearly created by humans, but it is unclear exactly when. How do we determine the age of a rock scratching—what sort of clock could we use? Here nuclear-dating methods fail us, since there is no way to carbon-date a figure etched in stone. Or is there?

It turns out that there is a way, and it comes in the form of an ancient insect infestation. Archaeologists studying the artwork at Kimberley noticed an abandoned wasp's nest covering one of the figures, which means that it must have been built after the figure was scratched into the rock.

Figure 1. *Rock art like this, discovered at Kimberley in northwestern Australia, provides clues about our distant ancestors.*

The nest could have been constructed a hundred years ago, of course, in which case it wouldn't have been terribly useful for dating the ancient art. But when they tested it, using a technique known as optical luminescence, they were surprised to learn that it was about 17,000 years old, which dates back to the Paleolithic. Moreover, the wasp's nest was built over a human figure with an elaborate headdress, which makes this the oldest human depiction in the world (the drawings in European caves are almost all animals). Richard Roberts, the researcher who tested the nest, thinks the drawings may be substantially older than this, since the climate in Kimberley 17,000 years ago was much drier and would not have been a likely place for human habitation.

But are the Kimberley drawings the oldest confirmed evidence of a human presence in Australia? No—farther south in New South Wales, near Sydney, human skeletons at Lake Mungo have been dated to between 45,000 and 50,000 years ago. This makes them the earliest modern humans ever found outside of Africa, predating the evidence in Asia by roughly 10,000 years. Is it possible that these people could have sailed directly from Africa to Australia, bypassing the Arabian Peninsula, India, and southeast Asia?

This conundrum has long puzzled archaeologists, with extreme views dominating the debate. Some have argued that modern Australians are the descendants of *Homo erectus* in southeast Asia, suggesting that they entered Australia 100,000 years ago or more. Others have theorized an influx of modern humans within the past 10,000 years. The truth lies somewhere in the middle, and genetics has garnered the clues necessary to answer the question.

Modern Australian Aborigines are not descended from *erectus* or other human cousins. Their DNA, while quite different from that found in people living in Asia due to their long period of isolation, is clearly part of the modern human lineage. Like all other non-African lineages, Australian Aborigines ultimately trace their ancestry back to Africa. So how did humans make it to Australia without apparently passing through Asia? To solve the riddle, we will need to leave aside both genetics and archaeology and turn once again to climatology.

WHEN ALL ELSE FAILS

As we saw in earlier chapters, the world's climate over the past 50,000 years has been rather unstable. Today we live in a benign, relatively warm period known as an interglacial, but in the past the world was a much colder place. While this had an obvious impact on the people living in the far north, those in the tropics were not spared the climatic upheaval.

The same effect that gave rise to the Bering land bridge and allowed Phil's ancestors to walk into the Americas from Siberia changed sea levels worldwide. A drop of 100 meters, or more than 300 feet, has been documented at various times during the last ice age because the water was tied up in the great ice sheets of the far north. While this amount may not seem like much (it's not much taller than a small skyscraper), the vertical difference is less important than what happened to the land area.

A quick glance at Figure 2 shows that Australia is no longer an island continent. Rather, it is joined to New Guinea as part of the ancient continent of Sundaland.

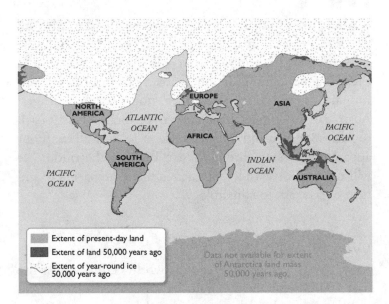

Figure 2. *A glacial maximum around 50,000 years ago meant that water was trapped at the polar ice caps, and coastlines were much lower as a result.*

This is because the Torres Strait separating the two countries is quite shallow, even though the currents are very strong. Even today Australia and New Guinea retain similar animal and plant species in the tropical zone, with oddities such as the velociraptor-like cassowary, a five-foot-tall predatory bird found in both northern Queensland and across the strait in the forests of New Guinea. The similar flora also reflects their ancient history as joined landmasses.

In the same way, other landmasses were much larger 50,000 years ago. The western coast of India would have been as much as 200 kilometers west of where it is today, and Sri Lanka would have been joined to it. Most of island Malaysia would have been joined to the peninsula.

Overall, the coast would have been very different. It is likely that this is the reason that there is no trace of modern humans along the early route to Australia. Their camps are underwater today, subsumed under the rising sea levels of the past 10,000 years. It suggests that the best place to look for evidence of an early human presence in Asia might not be inland, but rather along the near-shore ocean bottom. Intriguingly, the earliest evidence of modern humans in the Indian subcontinent has been found in a cave near the Sri Lankan shore, suggesting a treasure trove of Paleolithic stone tools may lie just under the water's surface.

But if the archaeological evidence fails to provide us with the clues we need to determine the route early Australians took, perhaps genetics can find the trail in the people living in Asia today.

THE NEEDLE IN A HAYSTACK

Humans migrating from Africa may have passed through India en route to populate the rest of southern Asia and Australia. Professor Pitchappan was not necessarily hoping that the Piramalai Kallar samples would reveal details about how Indians were related to Australians when he suggested our collaboration. Rather, he hoped to investigate the relationships among Indian populations, with their complex history of caste, conquest, and migration. But when we tested several hundred samples collected in southern India, we found a clear genetic link to Australia. The first piece of evidence came from Virumandi.

The genetic link between his Y chromosome and Australia came in the form of a marker known as RPS4Y (so called because it is found in the gene encoding **R**ibsomal **P**rotein **S4** on the **Y** chromosome) or simply M130—the 130th marker to be discovered on the Y chromosome. It is found today at a frequency of around 5 percent in southern Indian populations, including the Piramalai Kallar. Yet it is the predominant lineage (greater than 50 percent) in Australian Aborigines, based on the few dozen samples that have been tested. It is found at a frequency of around 20 percent in southeast Asia, and we have traced a clear genetic trail along the southern coast of Asia.

At the time I visited in 2002, Virumandi was a 26-year-old librarian at Madurai Kamaraj University. I had returned to the village to reveal the details of the genetic analyses we had performed on the other members of his village. We were welcomed into Virumandi's humble house, where he lived with his wife and parents, and shared a cup of tea. We then gathered the other members of the village to tell them the remarkable story of their connection to Australia. They were awestruck to have played such an important role in this scientific discovery. We shook hands and left, happy that we had been able to reveal the results to them in person.

The Piramalai Kallar carry a piece of the genetic trail—a link to the aboriginal Australians—in the same way that the Chukchi carry a genetic link to Phil and other Native Americans. M130 is found in northern India, but there it is a more recent, derived type with additional markers (part of a sublineage known as C3), which is also common in Central Asia and Mongolia. It probably originated in Mongolia and was spread, in part, by the Mongol conquests of the 13th century.

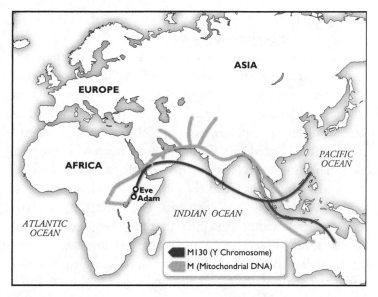

Figure 3. *The first humans to leave Africa likely followed the southern Indian coastline, reaching Australia in only a few thousand years.*

The lineage found in Virumandi and other southern Indians is directly ancestral to the same C lineage found in Australians. This tells us that C appeared first along the coastal route leading to Australia from Africa, and only later did it migrate inland to Mongolia, probably via the coast of East Asia (Figure 3). Haplogroup C3 seems to have continued its coastal journey all the way into the Americas, where some North American populations have fairly high frequencies of this lineage. These groups typically belong to those speaking the Na-Dene languages, the second major linguistic family in North America. From the age of the C3 lineages in North America it appears that the ancestors of the Na-Dene speakers entered the continent only in the past 8,000 years. At this time the

Bering land bridge was once again submerged, so this journey must have been made by boat.

The mitochondrial DNA shows a similar pattern, although it is less clearly defined. Macrohaplogroup M is largely distributed along the southern route and is not seen in the Middle East. However, it is found in India at a frequency of 40 to 50 percent and in southeast Asia at around 20 percent. Like M130, though, haplogroup M is the predominant lineage in Australian Aborigines. It appears that men and women made this long coastal voyage together, on foot, around 50,000 years ago. The patterns of diversity generated within Australia in the past 50,000 years, however, remain largely unknown due to the small number of Australian Aborigines that have been sampled. One of the main goals of the Genographic Project will be to learn more about these ancient patterns.

AN INLAND WAVE

The Indian and southeast Asian distributions clearly show that the descendants of these early coastal migrants are a minority today. Most people in India, and in fact, more than 95 percent of Y chromosomes and 50 percent of mitochondrial lineages in Eurasia, Australia, and the Americas, trace their ancestry back to a different migration—the one that led Phil's ancestors into Central Asia 40,000 years ago—an inland route through the Middle East. This new group of migrants would become one of the most influential in the spread of humans worldwide.

During the last ice age, when sea levels were dropping along the coastal route that led Virumandi's ancestors to India, inland conditions were also changing. Because of the generally drier conditions, the savannas of East Africa were expanding into areas that had formerly been forest. The Earth's climate was going through a period of erratic weather, with rapid increases in temperature interspersed among the generally lower temperatures. The result was that humans were pushed and pulled in many directions, their range alternately expanding and contracting. There is evidence that the Sahara was sometimes wetter than it is today, as well as drier (Figure 4). During the wetter phases savanna would have intruded into today's desert zone, allowing humans and the animals they hunted to move northward.

These early bands of hunting and gathering humans would have been drawn into the expanding grasslands by a wetter climate, but worsening conditions could also have pushed them out. The neurobiologist William Calvin, who has written on the effects of climate change on early human evolution, compares the Sahara during this period to a kind of pump, drawing in animals from other regions during wetter phases and expelling them when the weather turned drier.

During one of these outward-pumping phases a small group of hominids left Africa 45,000 to 50,000 years ago and entered the Middle East. The exact route they took is unknown, although current Genographic work on the populations of the Sahara—particularly those in Chad, Sudan, and Egypt—should shed some light on this. By 45,000 years ago, this small band was well established in the Middle East and left behind ample evidence in the form of skulls, tools, and other archaeological material. There is also a unique genetic marker called M89, the ancestor of much of the male population of the Northern

Hemisphere. M89 appears to have been accompanied on his journey through the grasslands by a woman belonging to the mitochondrial DNA clan N, who also left Africa for the Middle East around this time.

This was not the first time that humans had migrated to the Middle East. Skulls and other fossilized remains belonging to the *Homo sapiens,* dating to 100,000 years ago, have been found at Qafzeh and Skuhl caves. But then they suddenly vanish about 30,000 years later. There is a long gap in the fossil record between their disappearance and when humans reappear in the Middle East about 50,000 years ago.

Once they were established on the Asian landmass, though, these people spread rapidly. According to the genetic evidence, these inland populations moved quickly into Central Asia and India, with one line leading to Phil's ancestors via markers M9 and M45. M89 and haplogroup N's descendants also spread into India and reached East Asia soon after. Early humans probably took only 10,000 years to colonize the majority of the Asian landmass below the Arctic. This final frontier would have to wait until Phil's ancestors made their icy trek 20,000 years ago.

The inland clans—M89 and mtDNA-N—soon expanded across regions formerly inhabited only by the coastal clans defined by M130 and mtDNA-M. Why they were so successful is unknown—perhaps their populations had developed cultural attributes that gave them an advantage. Clearly, though, frequencies of the coastal lineages today are a fraction of their former levels, with almost all Indian men tracing their ancestry back to the incoming grassland populations.

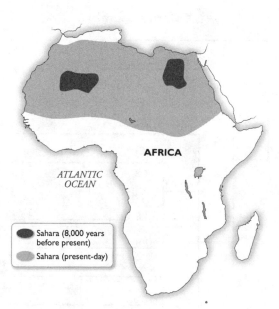

Figure 4. *Periodic fluctuations in climate have caused the Sahara to shrink considerably compared to today's wide distribution.*

EURASIAN ADAM AND EVE

Virumandi's defining marker, M130, does not occur in isolation—it also had ancestors. Tracing its roots brings us to a very important marker indeed—M168. It first occurred in a man who was born between 50,000 to 60,000 years ago, probably in northeastern Africa. This man gave rise to the great coastal exodus that led to India and on to Australia. And one of his great-, great- . . . grandsons, a few thousand years later, also gave rise to the M89 lineage that left Africa for the Middle East.

On the female side, a single mtDNA type—carried by one woman—gave rise to both of the M and N lineages found outside of Africa. Her lineage, L3a, began around 60,000 years ago, roughly contemporaneous with M168.

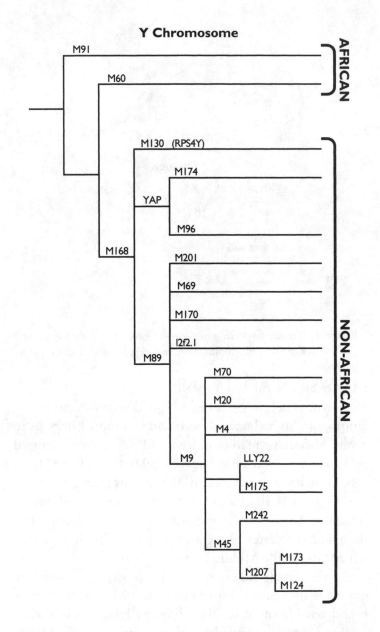

mitochondrial DNA

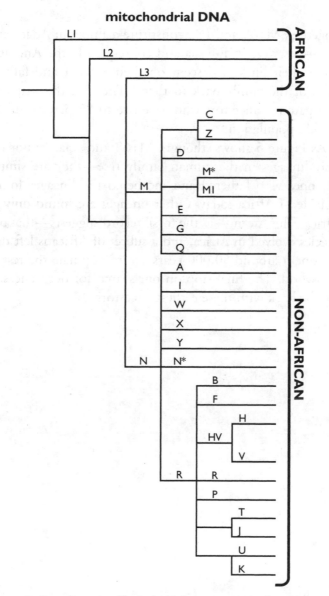

Figure 5. *Deep lineages of both the Y-chromosome mitochondrial trees, showing the African vs. non-African branches.*

Together, M168 and L3a constitute a Eurasian Adam and Eve—everyone in Eurasia and consequently the Americas traces their lines of descent on their mother and father's sides of the family back to them. They are the maternal and paternal ancestors who gave rise to 85 percent of the world's population.

As Figure 5 shows, though, M168 and L3a are not the only lineages in the human family tree—they are simply the ones who fathered and mothered the lineages found outside of Africa today. Other lineages are found only in Africa. This means—as the fossil record suggests—that our species evolved in Africa, and a subset of Africans left that continent around 50,000 years ago to populate the rest of the world. The final stop on our search for deep ancestry will be back with these earliest ancestors.

6

JULIUS'S STORY
The Cradle

I met Julius in January 2005 while making the film *The Search for Adam*. The film explained how we had used genetic data from our studies of the Y chromosome to discover a shared common ancestor—a man from whom all men today are descended. Julius was the chief of a tribe of Hadzabe, a group of hunter-gatherers who live in Tanzania at the edge of the great Rift Valley.

The Hadzabe, or Hadza, are a rarity in the modern world. Today, almost everyone is the descendant of someone who started to grow their food in the past 10,000 years. Agriculture is so pervasive that you could be forgiven for assuming that everyone practices it, just as many Western children think that beef and chicken only come cleanly wrapped in plastic and displayed in supermarket refrigerated cases. But until recently, of course, eating meat meant raising the animal and then slaughtering it yourself, rather than simply choosing the cut that looked freshest and paying with a credit card.

The Hadza are people who—like our distant ancestors of 50,000 years ago—survive by hunting down and killing wild animals, gathering wild plants, and locating natural sources of water. Spending time with the Hadzabe was like visiting the preagricultural era, before governments, empires, towns, and even villages existed.

So few true hunter-gatherers have survived into the 21st century that experiences like the one I had with the Hadza are extremely rare. Until recently most governments regarded them as an embarrassing anachronism to be channeled into heavy-handed social programs. Julius, for instance, had been forced to attend a local school as a child and was not allowed to speak his ancestral language. During a less enlightened era when hunter-gatherers were seen as impediments to economic development, the same sort of social engineering happened to the Aborigines in Australia, many Siberian tribes in the Soviet Union, and to the Inuit and other groups in the Americas. In part because of this, only tens of thousands of hunter-gatherers are left in the world, and even possibly fewer if a strict definition is applied.

Despite the recent marginalization by governments, the hunter-gatherer way of life is an extremely successful adaptation. For the entire span of hominid history—millions of years—until 10,000 years ago (and even more recently in the case of northwestern Europeans) all humans lived in this way. That Julius's small band of 20 or so people has maintained a way of life that the rest of us gave up long ago, is a testament to its long-term success, not a sign of backwardness.

While the way of life of hunter-gatherers such as the Hadzabe is an amazing window into the past, it does not mean that the people themselves are in any way more primitive than the surrounding agricultural populations. The language spoken by the Hadza, for instance, includes a complex set of percussive sounds—like a cork popping, or the "tsk tsk" sound English speakers make to show disapproval—as part of its word construction. Some related languages in its family (called the Khoisan) have more than a hundred different sounds, while English and most European languages have only around thirty sounds. The click languages may even be among the earliest languages spoken by our species.

CLICKING

Language is one of the defining features of humans. While other species communicate with simple, nonverbal cues, and some may even convey relatively simple ideas using vocalizations (birds and whales, for example), we are the only species that has ever evolved the ability to communicate complex ideas using long strings of words.

Verbal communication is one of the most difficult things we do. Think about a typical conversation, which not only requires mastery of the grammatical rules that underlie the structure of the language, knowledge of a wide variety of vocabulary terms as well as short- and long-term memories to provide context, but also a complex set of motor skills involving roughly a hundred muscles in the face, mouth, and throat. The latter is important, as young children typically understand far more words than they can pronounce correctly. The real excitement of our language ability, though, lies within our brains.

Figure 1. *Stone artifacts over the past one million years* (top) *show relatively little change until around 50,000 years ago, when complex technologies and the first evidence for art become widespread* (bottom).

Chimpanzees, which lack the finely tuned motor apparatus necessary to produce speech, can be taught sign language as a means of communication. While they can construct a wide variety of two-word sentences to express simple ideas, like "eat banana" or "go outside," they lack the syntax necessary to produce a complex sentence such as the one you are reading now. Because of this great gap between us and other species, even intelligent ones like chimps, anthropologists believe that language developed relatively late in hominid evolution.

According to genetic data, the lineages leading to chimpanzees and humans split around six million years ago.

These dates were arrived at through the same mechanism we learned about in Chapter Two, namely counting the number of nucleotide differences between human and chimp genes and making use of the known mutation rate to calculate how long they have been diverging from each other. Assuming that our first ancestor on the line leading to humans had language abilities no better than all other apes do today—almost certainly a valid assumption—this means that our language abilities must have developed in the past six million years. But when along the evolutionary path did that occur?

One way of answering this question is to examine the changing patterns of anatomy on the line leading to modern humans. While we are uncertain of exactly what our earliest ancestors looked like, they were more ape-like than we are. The first marked transition was the emergence of bipedalism, or walking upright on two legs. This happened long before the second major step forward—an enlarged brain—and could have arisen as early as 4.5 million years ago in the hominid *Ardipithecus*. Why hominids became bipedal is greatly debated, although rapid movement in the grassland environment where our early ancestors evolved was probably a contributing factor, as was reduced sun exposure during the heat of the day. Since brain size had not increased significantly, and tools are not in evidence from these early hominid sites, Darwin's theory that bipedalism freed the hands for tool use is an unlikely reason for them to walk upright.

The next major leap occurred when brains became bigger. Brain size leapt from an ape-like few hundred cubic centimeters in the australopithecines to 600 to 700 cubic

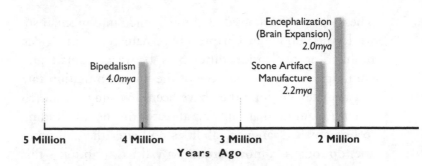

| 5 Million | 4 Million | 3 Million | 2 Million |

Years Ago

Bipedalism
4.0mya

Stone Artifact
Manufacture
2.2mya

Encephalization
(Brain Expansion)
2.0mya

centimeters in *Homo habilis*, the first member of our genus, and 800 to 1,200 cubic centimeters for *Homo erectus*. Hominids began to use tools during the time of *Homo habilis*, around 2.3 million years ago (Figure 1), and the increasing complexity of thought that this represented likely drove the increase in brain size.

Brain size continued to grow for the next 1.5 million years, until by 500,000 years ago it measured 1,300 cubic centimeters, comparable to that of modern humans (see timeline above, Figure 2). These big-brained hominids would have been the ancestors of Neandertals as well, and we now know that the Neandertals had on average 10 percent larger brains than modern humans. Size isn't everything, though, as the results indicate: We drove the Neandertals to extinction within a few thousand years of entering their stronghold in western Europe, even though our brains were not as massive.

So why did we win out against the Neandertals? We were smarter, and our improved brains probably grew in tandem with the development of fully modern, syntactic language, like that we use today, which brings us to the final stage in the development of modern humans. While earlier hominids probably had the physical capacity to

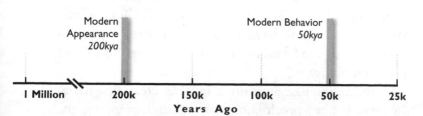

Figure 2. *Major advances in the development of modern humans over the past five million years*

speak and may have done so as well—although in less mellifluous ways than a Shakespearean sonnet—they almost certainly lacked the brain wiring to be able to say much that was worthwhile. Their speech was probably limited to simple chimp-like statements along the lines of "go there" or "eat now." The speech-enabling hyoid bones (the bone in your Adam's apple) found in Neandertal remains attest to a throat structure that would allow spoken language, but the other cultural attributes (such as the fact that the Neandertals' basic tools had not changed in a million years) suggest that our brawny cousins had little to say.

Human speech is part of a larger suite of characteristics that probably appeared about 50,000 years ago, according to anthropologists. The "Great Leap Forward," as biogeographer Jared Diamond calls the Upper Paleolithic transition in his book *The Third Chimpanzee*, represented a leap of understanding as opposed to anatomy. The difficulties faced during the last ice age probably led our ancestors of 50,000 years ago to develop better ways of exploiting the shrinking food resources in a harsh tropical environment.

Genetic data suggest that the human population size crashed to as few as 2,000 people around this time, and archaeological data is quite sparse as well. But coming back from the brink seemed to have involved a cultural transition unparalleled in our species' long history of upward mobility.

Modeling the past and future involves creating a narrative, much like speech. Our way of understanding the world seems to involve telling a story about it: John walked from the beach to the house. This simple sentence tells us all we need to know, about the locations, the sequence (beach first, then house), and the mode of action (walking). It even suggests that John is still in the house, which gives us a good clue if we want to find John now, all contained in one simple sentence. Try to convey all of this using two word sentences and you soon get bogged down in a confusing variety of alternatives. John beach. John house. John walk. Beach house. What does it all mean? So syntax, and the inherent narrative structure it provides, is a great adaptation if you want to explain complex events (imagine describing how to program a number into your cell phone using two-word sentences).

This new ability to explain complex events—which probably included where to find succulent roots or fruit, or how to hunt rare prey on the dry grasslands 50,000 years ago—would have provided an enormous advantage to our ancestors. So great was this advance that every human alive today, barring a severe medical disorder, has the ability to speak. The diversity of languages is astounding, with more than 6,000 spoken around the world today. All rely on syntax to convey complex thoughts, and the language you are currently reading is a direct descendant of those first tentative syntactic steps that took place 50,000 years ago. These early steps probably allowed the shrinking human population not only to survive their dire predicament, but also to thrive, expanding their range from Africa to encompass all six habitable continents.

This expansion appears to have occurred quite soon after the great leap in brain function. After the wave populating southern Asia and Australia, a slightly later one, exiting Africa via the Middle East, led to the populating of much of the Northern Hemisphere. And some people stayed on in Africa, including Julius's ancestors. The uniting factor among all of these seemingly disparate branches of humanity is complex, fully modern human behavior. We are all the descendants of a remarkable set of ancestors who set in motion the way we look, act, and think today. The next stop on our journey will be to use genetics to tell more about who these people were and where they might have lived in Africa.

JULIUS'S TEST

Julius agreed to be tested because he was curious about how science could tell him about his distant ancestors. As with most indigenous people, he had a clear sense of where his people had come from. According to their tradition, they had always lived in their homeland, with the baobab trees and the animals and the lake and Ngorongoro Crater rising like a gentle giant out of the African savanna. Although the incoming Maasai cattle herders—relative newcomers to the region, around 500 years ago—had displaced the Hadzabe from their former hunting grounds inside the crater, they still lived in essentially the same region where they had always lived.

When I explained what we were able to do by looking at his DNA, he took it all in with a wisdom that made him seem older than his 40 or so years (he wasn't certain about his birthday). He understood that he was carrying some sort of "essence" that connected him to his parents and grandparents, to the other Hadzabe in his tribe, and even to people in other parts of the world, and he wanted to see what we could find out. Perhaps more important, he also wanted us to tell his story so the world would know about his people, whose way of life was endangered by the encroaching modern world.

We swabbed his cheek and took it back to the laboratory for testing. The markers revealed that Julius's ancestors had indeed been living in Africa for a long time—in fact, since the earliest days of our species. His results revealed the presence of a Y-chromosome marker known as M60, which made him a part of haplogroup B. This group defines one of the deepest branches in the Y-chromosome tree, which separated from the others around 60,000 years ago, during the harsh early days of the last ice age.

We revealed his results to Julius when he came to National Geographic for the launch of the Project in April 2005. He was happy to share the story of his people as part of the launch festivities, and I was happy that I could tell him that his story about the Hadzabe origins really seemed to be true—that they had always lived where they live today and probably in eastern Africa. We hoped to learn much more about the origin and distribution of his ancient lineage over the coming years as the Project progressed; the results somehow bonded us—me as a scientist, him as the "owner" of the lineage—in a shared journey to discover more about the earliest days of our species.

The fact that his people had this lineage meant that they carried a glimpse of our most ancient ancestors in their DNA. Julius was doing us all an enormous favor by collaborating in our scientific effort.

After the launch, Julius and Phil Bluehouse visited New York. I had dinner with them a few days later and asked Julius what he thought of the city. "Very busy—people move too fast," he responded. I laughed and told him that I completely understood. When I had visited the Hadzabe in Tanzania several months earlier, it felt like a sort of homecoming. The peacefulness of their way of life was apparent, even in the face of modern challenges. The hunter-gatherer existence felt far more natural than our crazy world of office work, traffic, and urban alienation. When Julius visited Shea Stadium on his final day in New York, the management welcomed him by flashing his name on the electronic scoreboard. It was the least we could do for a true genetic VIP.

What can we say about the distribution of Julius's haplogroup, B? Mostly found in central and eastern Africa, it is common in the pygmies of the central African forests, as well as in the Hadzabe. But it is not the deepest lineage in the Y-chromosome tree (Figure 3). This honor falls to haplogroup A, which is found in Ethiopia and Sudan, as well as in the San Bushmen of southern Africa.

Interestingly, the San Bushmen of the Kalahari, like the Hadzabe, also speak click languages. This shared feature has led some linguists to suggest that the click speakers were once distributed throughout eastern Africa and became forced into their present distribution only with the expansion of the other larger African language families. The most important of these expansions was that of the

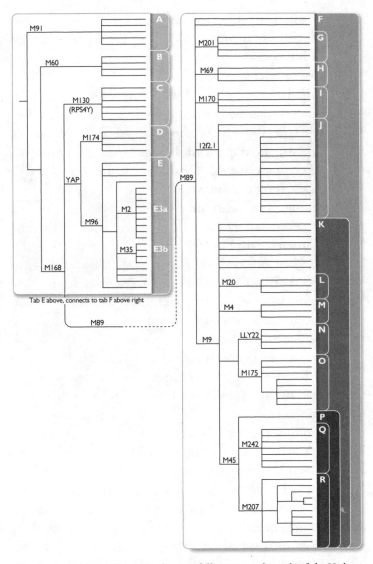

Figure 3. *Every man's genetic lineage falls into one branch of the Y-chromosome tree. Because of ancient migrations, genetic drift, and population bottlenecks, the deeper groupings exhibit some geographic specificity.*

Bantu language speakers, who probably originated in west-central Africa, in the region of present day Cameroon or southeastern Nigeria. The Bantu speakers migrated eastward and southward over the past 2,500 years as their farming techniques led to a population expansion, possibly aided by their mastery of ironworking. As they moved into other regions of Africa, the hunter-gatherer click-language speakers already living there appear to have been replaced by Bantu languages. Fossil evidence suggests that 10,000 years ago people similar in appearance to the modern San Bushmen were once found as far north as Ethiopia, indicating that their present distribution is the remnant of a once widespread people.

Haplogroup E3a has been associated with the spread of the Bantu speakers. E3a is closely related to E3b, and the two are united by the presence of a shared marker known as YAP, which stands for **Y** *Alu* **P**olymorphism. *Alu* elements are short pieces of DNA (around 300 nucleotides in length) that have been inserted into the mammalian genome in millions of random locations over the past 65 million years, in a kind of viral reproductive process. They comprise around 11 percent of the human genome, making up a substantial portion of so-called "junk DNA." The function of junk DNA remains unknown, only that it has been transmitted through the generations as harmless genetic baggage. The YAP insertion that defines haplogroup E arose on one man's Y chromosome around 50,000 years ago in Africa. Some of his descendants with an additional mutation, M96, later left Africa, creating the distribution of E3b we see today. Others stayed behind and had an additional change on their Y chromosomes, M2, which defines E3a. E3a individuals are today found throughout Africa, but the age and distribution of the lineage suggests that it was spread over the past 2,500 years by Bantu-speaking populations who originated in west-central Africa.

Figure 4 shows that the A and E3a lineages are found at quite a high frequency in many African populations. A is most common in northeastern and southwestern Africa, while E3a is found at high frequency in most populations, particularly those associated with Bantu languages. The third major African lineage is haplogroup B, which is distributed throughout Africa. It probably spread at the time modern humans first populated the African continent.

LINGUISTIC FOSSILS

Whether these eastern San-like people spoke click languages or not is unknown—in the absence of writing, languages don't leave fossils—but recent genetic research on the Hadzabe and the southern San has discovered evidence for an ancient separation of the two groups. The authors of this study, led by Alec Knight and Joanna Mountain at Stanford University, found that the San and the Hadzabe were characterized by quite different genetic lineages. Both groups had high frequencies of haplogroup E3a, consistent with substantial admixture with Bantu speakers. It is the deeper lineages in the tree that reveal more, though. The Hadzabe are predominantly haplogroup B, like Julius, while the San Y chromosomes are mostly haplogroup A.

Of course the different Y-chromosome lineages in the Hadzabe and the San may have been caused by genetic

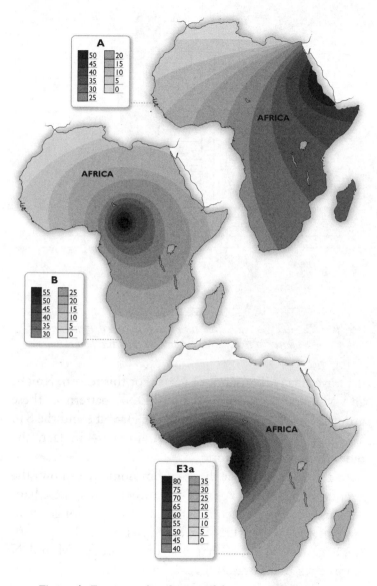

Figure 4. *Frequency distribution of the major Y-chromosome haplogroups in Africa: A, B, and E3a.*

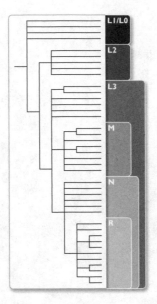

Figure 5. *Everyone's mitochondrial DNA falls into one line of this tree. The lineages group into six main branches that are likely representative of that person's maternal ancestral homeland.*

drift over many thousands of years. For this reason, Knight and Mountain also studied the mtDNA pattern in these groups. Their results showed that the Hadzabe and the San also had quite divergent mtDNA lineages—in fact, the oldest lineages in the world.

In the same way that the Y-chromosome tree shows the deepest split between A and B—in other words, these haplogroups have been accumulating diversity for longer than any others—the mtDNA tree shows a similar pattern (Figure 5). Remember that macrohaplogroups M and N led us back to a Eurasian Eve defined by the lineage L3. It turns out that there is additional diversity in L defined by lineages L0, L1, and L2. L0 and L1 are often grouped together as L1/L0 as the oldest lineage in the mtDNA tree, which separated more than 100,000 years ago. L2 is slightly more recent and probably arose in the past 80,000 years. All currently appear to be confined to African populations and together constitute the female counterparts to the Y-chromosome A and B lineages.

When Knight and Mountain compared the patterns in the Hadzabe and the San, they saw the same pattern they had seen in the Y-chromosome data. The two populations had very deep, but different, lineages. The San had high frequencies of L1/L0 lineages, while the Hadzabe had none of these—they were primarily haplogroup L2 (Figure 6). The similarity of the Y and mtDNA results is striking, and together the results suggest that the Hadzabe and San click language speakers are not merely the remnants of a once continuous population that was recently split, but that they had been living in their current locations—separate from one another—for tens of thousands of years. This suggests that their shared languages also date to a time prior to the recent Bantu expansion, and that click languages may be tens of thousands of years old. Using the estimated separation time of the Y and mtDNA lineages, these groups—and their languages—could be more than 50,000 years old.

The genetic data dates the age of that language family to the earliest days of demonstrably modern human behavior, at the dawn of the Upper Paleolithic. If click languages were spoken by the ancestors of the San and the Hadzabe, it suggests that they may have been among the earliest languages—and perhaps *the* earliest. DNA, then, has given us a tool to learn how our earliest ancestors communicated with each other. Could it also provide us with a clue as to what they looked like?

FACE-TO-FACE

Genetically, Africa is the most diverse continent in the world. Two Africans sampled from the same village could have Y-chromosome or mtDNA lineages that are more divergent from each other than either is to a non-African. This diversity extends also to physical appearance, where there is a broad range in different regions. The features North Americans and Europeans typically associate with Africans are influenced by the populations they have had contact with, notably those from west-central Africa during the slave-trading era. These people, speaking Bantu languages, have certainly had a large influence on the appearance of African populations, and today their genetic legacy is as widespread as their linguistic influence. But Africa contains a far wider range of appearances, as you can see in Figure 7.

The range of variation in Africa is extraordinary, containing both the tallest (the Maasai) and the shortest (Pygmies) people on Earth. Facial characteristics are similarly diverse. The heavy features of the West Africans contrast with the finer-boned features of the San Bushmen. The San also have an epicanthic fold—the extra layer of skin above the eyes that characterizes people from East Asia—while other African groups lack this feature. The skin coloration also varies enormously, ranging from a deep brown in the West Africans to the lighter skin of the San. Overall, the main uniting feature is a generally darker skin color for Africans relative to people from northern latitudes, such as Europeans.

Much has been made of skin color as a defining racial characteristic. Yet most human variation is found among individuals within populations, and less than 10 percent

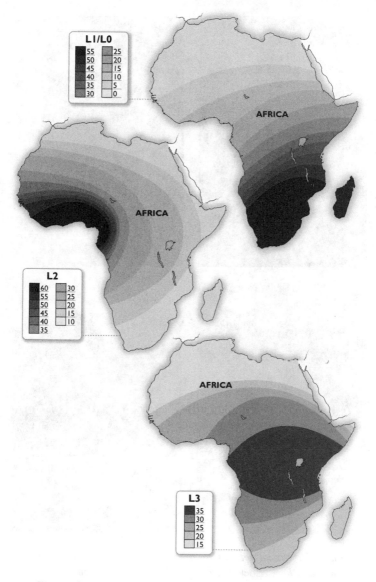

Figure 6. *Frequency distribution of the major mitochondrial haplogroups in Africa: L1/L0, L2, and L3.*

Figure 7. *African populations are the most genetically diverse on the planet. Above* (clockwise from top left) *are individuals from the Bantu of western Africa, the Maasai of eastern Africa, the San Bushman of southern Africa, and the Pygmys of central Africa.*

serves to distinguish between racial groups. In other words, beneath the skin we are all much more similar than our surface features would lead us to believe. The reason that our surface appearance varies so much is probably due to the survival of different genetic variants over time. Skin color has probably been subject to strong selection over the past 50,000 years, and the lighter skin of Europeans is most likely due to the climatic influences they encountered living far from their African homeland.

Africa is the most tropical continent in the world—the vast majority of it is located between the Tropics of Cancer and Capricorn. As any tourist on a tropical holiday can attest, the sun there is incredibly intense, resulting in painful sunburns for those whose skin is not used to it. The dark skin of Africans—and other tropical populations, such as southern Indians and Melanesians—is probably an adaptation to the sun's intensity. Since we are a relatively hairless species, much of our body surface is exposed to the scorching effects of the sun. Melanin, which makes skin dark, is a natural sunscreen, and since we evolved in the tropics, our early ancestors probably had dark skin to protect them from the sun's rays. As painful and debilitating as sunburn is, long-term exposure of light skin to the tropical sun can lead to skin cancer. Lighter skin also allows enough ultraviolet radiation to penetrate to the deeper skin layers and break down folic acid, which can lead to anemia and (in gestating fetuses) neural tube defects. Clearly it made sense for early humans living in the tropical zone to have darker skin. Consistent with this, the ancestral form of the major gene for skin coloration, known as melanocortin recaptor (MC1R), produces darker pigmentation.

As humans began to migrate out of the tropics, though, they encountered regions with much less UV light. When they started moving into these regions, they experienced a relaxation in the selection pressure needed to maintain the ancestral form of the MC1R gene, and once they moved far enough north the selection pressure reversed itself. This is because some UV light must penetrate the skin in order to biosynthesize vitamin D, an essential vitamin—so called because it has to be supplemented in the diet, or (in sunny conditions) synthesized. Without enough vitamin D, the children of dark-skinned people would have fallen prey to rickets, a nasty childhood disease in which the long bones are deformed due to an inadequate supply of calcium.

The need to allow more UV light to penetrate the skin would have favored mutations in MC1R that produced lighter skin. The function of MC1R can be reduced in several ways, and there are several different mutations that contribute in varying degrees to lighter skin. All would have been favored, though, during the cold northward trek of our ancestors.

Why has the marvelous adaptation become such an integral part of racist ideology, causing lighter-skinned Europeans to feel superior to the darker-skinned peoples of the tropics? It is probably due to the general correlation between latitude and levels of economic development over the past 2,000 years. As Jared Diamond has argued persuasively in his book *Guns, Germs, and Steel,* the peoples of the tropics have tended to be poorer than Eurasians during this period, and Africans have been among the poorest of the poor. What originated as an ice age adaptation to climatic diversity has been conflated with mental and cultural superiority, when the two phenomena have nothing to do with each other apart from geography. Again, the differences in skin color that distinguish human races are literally not more than skin deep.

If Africans have had dark skin for 50,000 years, it is a certainty that our distant ancestors were dark-skinned. They also were likely to have been thin people of a height similar to African hunter-gatherer populations today (perhaps 5'6" on average), since the stoutness of the Eurasian Neandertals was a biological adaptation to the climate evolved over hundreds of thousands of years. Since humans migrated into the northern parts of Eurasia relatively recently in comparison, they have had to adapt culturally for their African body proportions. Basically, though, our earliest ancestors of 50,000 years ago probably looked very much like modern Africans.

A LONG TIME TO WAIT

We've come to the end of our journey back to the earliest human ancestors who can be detected in our DNA. Each person alive today can be assigned to one of our haplogroup clans, uniting their Y chromosome or mtDNA with others who shared a similar journey to wherever they live today. But what about the root of the trees shown in Figures 3 and 5? What do these roots actually represent, and when did they live?

Any piece of DNA that is not shuffled through the action of recombination can be traced back in time to an earlier ancestor. There are 6.5 billion such pieces of mtDNA today and around half that number of Y chromosomes; all of them can all be traced back to a sole root.

This entity, known as the coalescence point, is the single mtDNA or Y-chromosome type from which they all trace their descent. In any given sample of nonrecombining DNA sequences there must be a single ancestor at some point in the past.

While this is a statistical certainty of DNA sequence evolution, it also has a certain resonance in this context since these coalescence points represent individual human beings who carried those pieces of DNA—a single woman whose mtDNA has been passed down to the present day by her wandering descendants, modified only by the mutational process. Similarly, the Y-chromosome coalescence point represents a single man who lived and died just like any other human. People essentially like us, and our earliest human ancestors are visible in our DNA.

The coalescence points have been given evocative names. The first to be discovered, that of mtDNA, was given the name Eve, representative of her role as the earliest female ancestor. When Rebecca Cann and her colleagues at the University of California, Berkeley, first described this ancestor in a landmark paper published in 1987, it made immediate news, including a famous article in the magazine *Newsweek* that showed an African-looking couple standing under the tree of knowledge holding an apple, with a knowing serpent looking on.

Despite the obvious comparison to the biblical story of Adam and Eve, the research shocked many, since Cann's team had not set out to find an African origin. Although the field of paleoanthropology had a long line of skeletal remains and tools providing a direct link between all modern humans and Africa, the fact that genetics—a science based on the accumulation of a large quantity of seemingly unrelated data—had arrived at that result based on DNA evidence surprised many, including members of the anthropological community. But the real shock came with the date.

Eve, according to Cann and her colleagues, lived less than 200,000 years ago, which means that all of the diversity we see in modern human populations—the range of skin colors, hair types, and skeletal features—has arisen since then. While the field of anthropology had no particular sense of how long it had taken to generate this physical diversity, most of them thought that it had taken a lot longer than 200,000 years. Carleton Coon, as you will recall from Chapter One, suggested that modern humans had evolved from ancestral hominids over the course of the past million years or so. Cann cut the amount of time allotted for this diversification down to a small fraction of this. Her research suggested that we all had an African great-, great- . . . grandmother who lived much more recently than we ever suspected. While Darwin had suggested that we all shared an ape-like ancestor in Africa, perhaps millions of years ago, no one had ever made such a bold assertion as the "young turk" geneticists.

The debate on how to interpret the Berkeley team's discovery went on for some time, with many attacking its inference of an African Eve (it was equally likely, it turned out at that time, to have had an Asian coalescence point), but the results were finally confirmed in a study published in 2000 that narrowed the estimated date to around 170,000 years. Mitochondrial Eve was African.

The field of genetics then turned to other pieces of DNA. While every stretch of the human genome continued to point toward an African root on the basis of the pattern of greater diversity in that continent, the real hope lay in the Y chromosome. If we could find an African Adam to match with our Eve, it would mean that the story of her origin must be true. The search for Adam, though, was far more difficult. The reason that mtDNA was the first piece of DNA to be applied to the study of human diversity was that it was relatively easy to work with. Today sequencing DNA is a simple exercise; even high school students can do it in their biology labs at school. What many forget is that DNA sequencing won its discoverers the Nobel Prize in 1980. Back then the methods used for sequencing were still labor-intensive, involving days of tedious biochemistry and radiation exposure to obtain a small region of sequence. For this reason, researchers looked at regions of the genome that would yield a large number of polymorphisms, or genetic markers, to study.

The problem with the Y chromosome was that it didn't vary that much from person to person. One of the early studies, in the early 1990s (when sequencing had become a bit easier) found no variation among 38 men chosen to represent world diversity. But the region they studied, 729 nucleotides in length, was relatively short. The field would have to wait until new techniques were devised for rapidly sequencing DNA.

These methods arrived in the mid-1990s when Peter Oefner and Peter Underhill, while working in Luca Cavalli-Sforza's laboratory at Stanford University, invented a new method for assessing sequence variation that did not rely on direct sequencing. Using it, the two discovered scores of new Y-chromosome polymorphisms. As a postdoctoral fellow, at the time I was amazed at the sensitivity of their technique and rapidly dropped other work to get involved. After all, who would turn down a chance to find Adam?

In a paper published in 2000, more than 13 years after the discovery of Eve, Underhill, Oefner, and 21 others (myself included) described the clearest view yet obtained of our common male ancestor. The study presented a genealogical tree—the basis for the Y-chromosome trees throughout this book—based on the new collection of Y-chromosome polymorphisms. It gave us a first glimpse at the fascinating geographic patterns we have explored in this book. It also showed how recently our common Y-chromosome ancestor lived—only 60,000 years ago.

The date was stunningly recent, since it revealed that all of the Y-chromosome diversity in the world had been generated in that comparatively short amount of time. We were surprised because it differed so much from the date for Eve. If she lived 170,000 years ago, and Adam lived 60,000 years ago, that's a long time to wait around for your mate to show up. Where were the men 170,000 years ago?

The reason we don't find male lineages coalescing at the same time as Eve is because of early human sexual behavior. In most traditional societies, a few men do most of the mating—think of chieftains and warlords, for instance. Some men never get to have children, while others have more than their fair share. This is known as the *variance in reproductive success,* and it is higher for men than women, which means that women have more equal opportunities to have children. Since women are passing on their mtDNA, the result is that—in general—mtDNA lineages have a more equal chance of being passed on to the next generation than Y-chromosome lineages. This behavioral quirk tends to reduce the *effective* population size of the Y chromosome, since not every man will pass on his Y, while most women will pass on their mtDNA. Since genetic drift acts more quickly in small populations, changes in the lineage composition of the Y-chromosome pool occurred more rapidly. Over time, this meant that Y lineages were more likely to be lost than mtDNA lineages. The result is that the deeper Y lineages were lost over the past 170,000 years, and the only ones left date to around 60,000 years ago.

While the difference between the Y and mtDNA dates are important technical issues, their only real significance for our story lies in placing the origin of modern humans in Africa. The more recent date—that of the Y-chromosome coalescence—tells us that modern humans were still living in Africa until 60,000 years ago, and only after this time did they leave the continent to populate the rest of the world. We made our journey around the world and generated the dizzying amount of diversity that defines us today—colors, sizes, languages, and cultures. This adds up to only 2,000 generations, which is the blink of an eye on an evolutionary timescale. Such a date also helps to make sense of the earlier results, such as Lewontin's finding that we all appear to be part of an extended family at the genetic level. If we only started diverging so recently, it's no wonder.

The complex tapestry of human diversity is woven together by genetic threads, connecting us through the migrations of our ancestors. Deciphering the complex patterns in this tapestry is the goal of the Genographic Project.

The end result will, we hope, serve to unite the world's people while respecting the incredible diversity that defines us as a species. If we can achieve this, the Project will have succeeded.

EPILOGUE
THE FUTURE OF THE GENOGRAPHIC PROJECT

I n this book we've taken a scientific journey back in time, through genealogy and history, back to prehistory and the earliest days of our species. Along the way we have picked up the key scientific concepts, learned about how we use DNA as a tool to study the past, and met the great-, great- . . . grandparents of every human alive today. So what's next?

The story told in this book is just a brief overview of where the Genographic Project is today. As I wrote in *The Journey of Man,* we have a glimpse of the forest, but we know very little about the trees. Much remains to be discovered, and the field today is just beginning to uncover the details that flesh out the human story. The Genographic Project will continue to explore that story and to add more knowledge to the human journey. To do this, however, we need far more samples than we have access to at the moment.

The story told in this book is based on approximately 10,000 samples that have been examined for a handful of Y-chromosome and mtDNA markers. Ten thousand people, out of a world population of 6.5 billion, simply is not a representative sample of humanity's diversity. Moreover, most of the samples that have been studied come from the places that are more accessible to scientists—North America, Europe, and East Asia. Africa remains terribly under-studied genetically, as do vast parts of Asia, Oceania, and the Americas. There are many stories waiting to be told, but we need the samples to tell them properly. It's not simply a question of pumping up the numbers, but rather of conducting focused sampling to solve historical riddles.

The research goals of such a project are ambitious, ranging from the most ancient questions about our origins in Africa to the impact of relatively recent events on genetic diversity, such as the reign of Genghis Khan or the conquests of Alexander the Great. The Genographic Project also aims to make some progress toward understanding the reasons for our tremendous phenotypic diversity, unusual among species of large primates. What role did natural selection, genetic drift, and sexual selection—choosing features we find attractive in our mates—play in molding our physical appearance?

THE TEAM

A project this large can only be carried out by a dedicated team, and the Genographic Project, guided by an ethical framework that adheres to the highest standards established by the scientific community, has assembled some of the best human population geneticists in the world. In North America, for instance, Theodore Schurr at the University of

Pennsylvania is focusing on the indigenous populations of Canada, the United States, Mexico, and the Caribbean, trying to answer questions about the settlement of the New World. Tad has spent much of the past two decades studying human populations in the Americas and Siberia, and his efforts in the project will help to shed light on questions such as:

• How many waves of migration were there to the Americas, and was one of the earliest migrations along the west coast?
• Could Europeans have migrated to the Americas thousands of years ago?
• Is there a genetic signal from the expansion of indigenous American agriculture—i.e., did the farmers or their culture move?

Fabricio R. Santos at the University of Minas Gerais in Brazil is heading up the South American regional center, focusing on the earliest days of human habitation on the continent and trying to explain its tremendous linguistic diversity. While acknowledging the sensitivities of working in his home region, Fabricio has previously carried out research on populations around the world, focusing his attention on the Y chromosome and what it reveals about human migratory patterns. His work will shed light on many outstanding issues in South American history:

• Were there any migrations to South America from the Pacific?
• Did the Inca Empire have a genetic impact on northwestern South America?

- How do we account for the extraordinary linguistic diversity found in native South Americans? Have populations there been separate for a long period of time?
- Can we find genetic signals of now extinct groups, like the Arawak in the Caribbean, in today's admixed populations?

Jaume Bertranpetit and David Comas, both at Pompeu Fabra University in Barcelona, Spain, and Lluis Quintana-Murci, at the Pasteur Institute in Paris, France, are focusing on genetic variation in European populations, using our already large sample sizes in this region to answer very specific questions about the history of Europe.

- Was there any mingling between modern humans and Neandertals during the European Upper Paleolithic?
- What route(s) did modern humans take in their settlement of Europe?
- Did the Celtic expansion of the mid-first millennium B.C. leave a genetic trail?
- Can we detect historical migrations during the past 2,000 years in today's European populations—Normans in Sicily, or the spread of Finno-Ugric languages in Eastern Europe?

Pierre A. Zalloua, at the American University of Beirut in Lebanon, is sampling in the Middle East and North Africa, the cradle of the Neolithic Revolution and homeland to some of history's greatest empires. Pierre has devoted much of the past decade to understanding Lebanon's genetic structure, and now he is turning his keen attention to broader questions about the history of the region, such as:

- Did important imperial conquests have an impact on the genetic landscape of the conquered regions? For instance, did Alexander the Great's armies leave a genetic trail?
- Where did the Afro-Asiatic languages, including Arabic and Hebrew, originate?
- Who were the aboriginal inhabitants of North Africa, and are the Berbers their direct descendants?
- How much genetic exchange has there been across the Sahara?

Himla Soodyall at the University of Witwatersrand in Johannesburg, South Africa, is taking on the challenge of sampling in sub-Saharan Africa, the most genetically diverse part of the world. Among other questions, she hopes to address:

- Which African populations harbor the most ancient genetic lineages, perhaps suggesting a geographic origin for modern humans?
- How has European colonialism had an impact on the genetic patterns in Africa?
- Can we trace the origins of the Bantu people—and their expansion across Africa—from genetic patterns?
- Was there a separate domestication of cattle in Africa? Did this lead to a population expansion?

Ramasamy Pitchappan at Madurai Kamaraj University is leading our regional center in India, the oldest continuously inhabited region outside of Africa. Its 400-plus languages and complex social system make sampling a tremendously challenging proposition.

- Where did the Dravidian speakers originate? Were they the "first" Indians?
- What role has the Indian caste system had in determining genetic patterns?
- Where did Indo-European speakers originate, and what languages were being spoken prior to their spread in Europe and Asia?

Elena Balanovska at the Russian Academy of Medical Sciences is covering a region that stretches across ten time zones, from the northern border of Finland to central Asia to the Bering Strait. It encompasses most of the former Soviet Union, with its incredible mix of ethnic and linguistic groups, and spans the entire continent of Asia.

- When and where did modern humans first colonize the Arctic?
- Who were the first inhabitants of the Caucasus region, and why is there such incredible linguistic diversity there?
- What role did the Silk Road play in dispersing genetic lineages?

Li Jin, our principal investigator for East Asia, is sampling from China to New Guinea from his base at Fudan University in Shanghai. He has already studied the populations of China in great detail, and is focusing his attentions on a large and complex region that holds keys to understanding the settlement of many other areas of the world, including Australia, Polynesia and potentially the Americas.

- How has the geography of China molded genetic patterns there?

- Who were the aboriginal inhabitants of Indonesia? Was there much genetic exchange with Australia?
- Was there any mingling with *Homo erectus* as modern humans spread throughout Southeast Asia?
- What are the patterns of genetic variation in New Guinea? Do they parallel the extraordinary linguistic diversity there?

Robert John Mitchell, based at La Trobe University in Australia, is working with populations in Australia and New Zealand, focusing on the migrations of their ancestors to their ancient homelands.

- How do the genetic patterns in Australia correlate with the Aboriginal songlines—their own oral histories?
- Can we use genetics to trace the spread of the Polynesians from island to island in the Pacific?

Finally, Alan Cooper at the University of Adelaide is adding a crucial component to the project in his studies of ancient DNA—a method that occasionally allows us a direct look at the genome of a long-dead individual. Cooper's team has developed many exciting new techniques for analyzing ancient DNA, and his work holds tremendous promise for our ability to answer key questions such as:

- Is it possible to obtain intact DNA from the remains of *Homo erectus* and other extinct hominid species, allowing us to resolve many debates in the field of paleoanthropology and early human history?
- Can ancient DNA be used to observe changes in the genetic composition of a population over time?

The Genographic Project team is fortunate to have the continued guidance of an advisory board comprising key disciplines—ethicists, indigenous advocates, paleontologists, anthropologists, linguists, and archaeologists.

THE CHALLENGES AHEAD

As of this writing (May 2006) the team has started sampling in places as far afield as Alaska, Chad, the Caucasus, southern India, and Laos. The coming years, before the estimated project completion date of 2010, promise to hold many surprises and much excitement. One of the most challenging aspects of the Project is coordinating the international efforts with indigenous and traditional peoples and their representatives to collect and evaluate their samples.

Earlier sampling efforts by others were largely haphazard affairs, in which a researcher typically collected a few samples for a particular study, then perhaps shared them with others interested in looking at different genetic markers. Another approach was suggested by Luca Cavalli-Sforza and his colleagues in the early 1990s in what is generally accepted as a landmark letter published in the scientific journal *Genome Research*. In their letter the researchers suggested a global sampling of human genetic diversity, a project that became known as the Human Genome Diversity Project, or HGDP.

The HGDP, while proposed with the best of scientific intentions, soon ran into considerable opposition from some indigenous groups. In large part this was because of fear about the negative potential of genetics, more prevalent in the zeitgeist of the nineties. In the era of Michael

Crichton's novel *Jurassic Park*, people were wary of how their DNA would be used. The HGDP's organizers didn't help their case by suggesting that genetic data, particularly those markers that were medically relevant and could be used to develop new drugs, could potentially be patented, thus enriching the researchers and their institutions, and the HGDP's focus on producing immortalized cell lines from blood samples—which would provide a living source of DNA in perpetuity, even after the donor was dead—ran contrary to many indigenous beliefs about the sanctity of the body. Although a few samples were collected under the auspices of the HGDP, for the most part it had closed shop by the late 1990s, killed by opposition to its many flaws.

The Genographic Project has a completely different approach. It is a purely anthropological, nonmedical, nonpolitical, nongovernmental, nonprofit international research project involving scientists from both the developed and developing world. Our goal is to enable indigenous communities rather than to take from them. We have gone to great lengths, from inception of the Project, to ensure that it is being carried out with the greatest transparency and in compliance with all local and international ethical and legal regulations for research involving human beings. Everyone who participates is doing so voluntarily—with the standard for consent in all cases being that it be free, prior, and informed.

We are consulting with indigenous peoples and advocates for indigenous rights as the project progresses to assure that we are addressing questions that are of interest to them in a culturally sensitive way. Sampling is tailored to take into account the wishes of communities or individuals (as culturally appropriate). In keeping with this commitment, our advisory board contains representation from indigenous groups and we have ongoing dialogues with indigenous advocates. Furthermore, our centers have been set up so that we are building capacity in the places where indigenous communities live. Also, our work does not focus on genomic regions with known medical relevance, thereby removing the temptation for other researchers or companies to profit from our work. Finally, we have said since the project launch that we see genetic discoveries as part of the common heritage of our species, and that no genetic data will ever be patented from the project. The Project is a collaborative effort between people from around the world who are interested in learning more about our shared history.

A great example of this collaboration in action happened in October and November 2005 when we ventured to Chad. Led by myself and Pierre Zalloua, the expedition would study populations who had never been included in previous work on African diversity. Chad occupies a unique position in Africa, one which the French refer to as a *carrefour* (or crossroads), since the country spans a zone that extends from the scorching Sahara in the north to the dense central African forests in the south, and sits astride the Sahel/ savanna zone that has connected the populations farther east—including those of the Nile region and the Rift Valley—with those living in West Africa. Its unique position suggests that the people living there may retain genetic clues about how these regions were settled by early humans.

Independence from France in 1960 left the country in a perilous state, however, with the Islamic north in con-flict with the largely Christian south. As a result, Chad's history is filled with civil wars and coups that has kept it locked away from the scientific mainstream for decades. The conflict in Darfur, on its border with Sudan, is just one in a long series of disturbances, and it is terra incognita genetically, which made it an exciting place for the Project to explore.

Our interest in the people of Chad was welcomed by the Chadian ambassador in Washington, D.C., and we launched ourselves into months of legwork in preparation for the trip. This included conversations with rebel leaders in the northern Tibesti region, who were fascinated by the Project and were keen to find out more about the deep ancestry of their own people. The years of war between north and south had isolated them, and they wanted to learn more about their people's stories—both genetic and otherwise—and to share them with the rest of the world.

Our first two weeks on the ground in Chad were spent preparing official collaborative agreements with our local scientific collaborators. A lack of a reliable telecommunications infrastructure meant that much of this work had to be done on site. But the long hours of work and preparations paid off when we were finally able to get out in the field and meet the people themselves. Driving along dusty, rutted roads that hadn't seen foreigners for many years, we wondered how we would be received by some of the more remote communities. Would they be receptive? Would they be interested in hearing about the Project?

Thankfully, these worries soon vanished. It was an incredible relief to see how warmly we were welcomed by these communities, all of whom wanted to discover more about their past. To a one, everyone we sampled wanted to know their personal results. They had heard stories about their ancestry—links to Nigeria, or Egypt, or even to

Jewish populations in the Middle East—and wanted to see what their DNA could reveal. One of the local leaders even described our efforts as "a gift from God"—a strong endorsement indeed. It was an amazing expedition, and the 300 samples we collected are already revealing many fascinating genetic patterns.

The Genographic Project is about more than basic scientific research, of course. Cognizant of the challenges faced by indigenous cultures, we intend to give something tangible back to the indigenous peoples themselves. The funds for this work, focused on cultural revitalization and education, come from the sale of the Genographic Project participant kits. As a nonprofit enterprise, National Geographic will put the net proceeds from the sale of these kits back into the project. Some will be spent on the field research, but the majority will go to help indigenous peoples around the world in the form of grants from the Genographic Legacy Fund. Projects will focus on cultural and educational initiatives, especially those proposed by the communities themselves. Our goal is to empower the indigenous peoples to tell their stories, to help revitalize their cultures, and to raise awareness about issues facing them.

We hope that the genetic results obtained from indigenous and traditional peoples, as with those obtained from public participants, will enrich everyone's sense of their own history. The genetic story is not meant to replace group beliefs or oral traditions but to complement them, leading to a fuller sense of where we all came from. The goal is ultimately to connect people from around the world into one family, showing how our ancestors took their long journey from Africa to wherever we live today.

APPENDIX
MITOCHONDRIAL DNA AND Y-CHROMOSOME
HAPLOGROUP DESCRIPTIONS

A ll humans belong to a haplogroup, an ancestral clan whose markers allow geneticists to study how humans came to inhabit the world. People can be sorted into haplogroups using their genetic patterns, and this appendix lists current descriptions for both mtDNA and Y-chromosome clans. (The latter starts on page 204.) Each description begins with the haplogroup's line of descent, starting from the earliest known common ancestor and moving down through the various branches of the family tree, listing the accumulation of genetic markers that define the haplogroup. A description of the geographic journey of the haplogroup follows, based on current knowledge about its age and distribution.

You can find a map of each haplogroup on the Genographic Project's Web site. The maps depict the inferred migratory path of the lineages from an African birthplace to where they are predominantly found in the world today. Simply go to *www.nationalgeographic.com/genographic/atlas.html* and click on "Genetic Markers" toward the bottom of the page. Select whether you want mtDNA or Y chromosome, then click on the haplogroup you are reading. These maps will be revised periodically to reflect the most up-to-date information.

As the Genographic Project progresses, we hope to learn more about the haplogroups described here, allowing us to refine our knowledge of how humans moved throughout the world all those years ago.

mtDNA HAPLOGROUPS

HAPLOGROUP *L1/L0*
Ancestral line: "Eve" → *L1 / L0*

Archaeological and fossil evidence suggests that humans originated in Africa during the middle Stone Age about 200,000 years ago. It is not until 50,000 to 70,000 years ago, however, with the onset of the late Stone Age (or Paleolithic) that signs of modern human behavior began to emerge.

Mitochondrial Eve represents the earliest female root of the human family tree. Her descendants, moving around within Africa, eventually split into two distinct groups, characterized by a different set of mutations their members carry. The older group is referred to as *L0.* These individuals have the most divergent genetic sequences of anybody alive today, which means they represent the oldest branches of the mitochondrial tree. Importantly, current genetic data indicates that indigenous people belonging to these groups are found exclusively in Africa, which supports the notion that the earliest humans originated in Africa.

The descendants of Mitochondrial Eve eventually formed another group called *L1,* which coexisted with *L0* individuals during these early years. *L1* descendants would eventually leave Africa and populate the rest of the world, while the descendants of *L0* remained in Africa.

Haplogroup *L0* likely originated in East Africa around 100,000 years ago. Over the course of tens of thousands of years, these early ancestors migrated throughout much of sub-Saharan Africa, and today are found at highest frequencies among geographically diverse populations such as the central African Pygmies and the Khoisan of southern Africa. Today, haplogroup *L0* is found in 20 to 25 percent of people in central, eastern, and southeastern Africa, but is found at lower frequencies in northern and western Africa.

The great Bantu migrations of the first millennium B.C. brought the Iron Age from West Africa to the rest of the continent. In the process, the indigenous *L0* individuals were likely assimilated or displaced, which helps explain why the descendant groups of *L1* are found at higher frequencies in many parts of central and eastern Africa.

Only recently have these ancient lineages made it out of the African continent, owing largely to the Atlantic slave trade. Many *L0* individuals in America share mitochondrial lineages with individuals from Mozambique. Overall, the continental frequency profiles of North and Central America are strikingly similar to those of western and west-central Africa, confirming that these were the source regions for American individuals with African ancestry. South America has a markedly higher proportion of *L0* individuals who derive their maternal ancestry from both west-central and southeastern African mtDNA lineages.

Some *L1* lineages are found at high frequencies in West Africa, calling into question the geographic homeland of this ancient group. However, because *L1* exhibits the greatest diversity, and thus age, in central and eastern Africa, it is likely that the high frequencies in West Africa are the result of a past reduction in population size. *L1* individuals can also be found in small percentages in some Arab populations: Palestinians, Jordanians, Syrians, Iraqis, and Bedouin.

HAPLOGROUP *L2*
Ancestral line: "Eve" → *L1/L0* → *L2*

L2 individuals are found in sub-Saharan Africa, and like their *L0/L1* predecessors, they also live in central Africa and as far away as South Africa. But whereas *L0/L1* individuals remained predominantly in eastern and southern Africa, *L2* ancestors expanded into other areas.

L2 individuals are the most frequent and widespread mtDNA haplogroup in Africa. The haplogroup can be separated into four unique subsets *(L2a, b, c,* and *d)*. *L2a* shows a wide distribution and is found at high frequencies in southeast Africa, but *L2b, L2c,* and *L2d* individuals are largely confined to western and western-central Africa. *L2d* is the oldest, with *L2b* and *L2c* splitting off most recently. It is therefore likely that the *L2* group, which arose around 70,000 years ago from a single female ancestor, first emerged in western and west-central Africa.

L2 is considered the signature Bantu haplogroup, accounting for roughly half of the genetic lineages found in the southeast African Bantu populations. The great Bantu migrations of the first millennium B.C. help explain why more recent *L2* lineages, descendants of *L1,* are found at high frequencies in many parts of Central and East Africa despite having arisen elsewhere.

Because of their geographic spread, particularly throughout West Africa, *L2* is also the most predominant African-American mitochondrial lineage, found at around 20 percent in the Americas. While some American lineages are shared with eastern and southeastern Africans, they are also present in West Africa. Unfortunately, the wide distribution of *L2* in Africa makes identifying precise geographic origins of the lineages difficult. Fortunately, current sampling with indigenous groups will help reveal more about those who made those transatlantic voyages hundreds of years ago.

HAPLOGROUP *L3*
Ancestral line: "Eve" → *L1/L0* → *L2* → *L3*

The most recent common ancestor of haplogroup *L3* is a woman who lived around 80,000 years ago. While *L3* individuals are found all over Africa, *L3* is particularly known for its northerly movements. They were the first modern humans to have left Africa, representing the deepest branches of the tree found outside of that continent.

Why would humans have first ventured out of the familiar African hunting grounds into unexplored lands? It is likely that a fluctuation in climate may have caused the *L3* exodus out of Africa. Around 50,000 years ago ice sheets in northern Europe began to melt, introducing a period of warmer temperatures and moister climate in Africa. Parts of the inhospitable Sahara briefly became habitable. As the drought-ridden desert changed to savanna, the animals expanded their range and began moving through the newly emerging green corridor of grasslands. The *L3* followed the good weather and plentiful game northward, although the exact route they followed is still being determined.

Today, *L3* individuals are found at high frequencies in populations across North Africa. From there, members of this group went in a few different directions. Some lineages within *L3* in the mid-Holocene headed south and are predominant in many Bantu groups found all over Africa. One group of individuals headed west and is primarily restricted to Atlantic western Africa, including the islands of Cabo Verde.

Other *L3* individuals kept moving northward, eventually leaving the African continent completely. Descendants of these people currently make up around 10 percent of the Middle Eastern population, and gave rise to two important haplogroups that went on to populate the rest of the world.

L3b and *L3d* individuals are mainly West African, with a few types shared with eastern and southeastern Africans. Occasionally *L3b* and *L3d* individuals can be found in southern Africa, though these are largely the result of the Bantu migrations that brought derived lineages into these ancestral southern territories.

In addition, *L3* is an important haplogroup because it is also found among many Americans of African ancestry. Most of the *L3* lineages found in the Americas are of West African origin, with some lineages representative of west-central and southeastern Africa also being found at lower frequencies.

Another subset of the *L3* haplogroup, *L3** individuals share some of the more common and widespread African-American mtDNA lineages, and most likely derive from West Africa between Angola and Cameroon, or from southeastern Africa in Mozambique. These lineages are also found among the São Tomé and Bioko in the Gulf of Guinea in West Africa. However, the transfer of slaves from Angola and Cameroon in the 18th century to work the sugar plantations of West Africa, and the subsequent transfer of slaves to the Americas, is responsible for the wide geographic distribution of *L3*s in western Africa and the Americas. It makes pinpointing their exact geographical origins difficult. The Genographic Project is working to better understand the origin and present distribution of these *L3* lineages.

HAPLOGROUP *M*
Ancestral line: "Eve" → *L1/L0* → *L2* → *L3* → *M*

Haplogroup *M* descended from *L3* and also left the African continent. These people most likely moved across the Horn of Africa, where a narrow span of water between the Red Sea and the Gulf of Aden separates the East African coastline from the Arabian Peninsula at Bab-el-Mandeb. The short ten miles would have been easily navigable for humans possessing early maritime technologies. This crossing constituted the start of a long coastal migration across the Middle East and southern Eurasia, eventually reaching all the way to Australia and Polynesia.

Haplogroup *M* is considered an Asian lineage, as it is found at high frequencies east of the Arabian Peninsula. Members of this group are virtually absent in the Levant (a coastal region in what is now Lebanon), though they are present at higher frequencies in the southern Arabian Peninsula at around 15 percent. Because the group's age is estimated at around 60,000 years old, members were likely among the first humans to leave Africa. Haplogroup *M* is found in East Africa, though at much lower frequencies than its subgroup *M1*.

This haplogroup is prevalent among populations living in the southern parts of Pakistan and northwest India, where it constitutes around 30 to 50 percent of the mitochondrial gene pool. The wide distribution and greater genetic diversity east of the Indus Valley indicates that these haplogroup *M*-bearing individuals are descended from the first inhabitants of southwestern Asia. These people underwent important expansions during the Paleolithic, and the fact that some East Asian haplogroup *M* lineages match those found in Central Asia indicates much more recent migration into Central Asia from the east.

Haplogroup *M* has several subbranches that exhibit some geographic specificity. Subgroups *M2-M6* are characteristic Indian subgroups. Haplogroup *M7* is distributed across the southern part of East Asia, and two of its own daughter groups, *M7a* and *M7b2,* are representative of Japanese and Korean populations, respectively. *M7* individuals reach frequency in southern China and Japan of around 15 percent, and are found at lower frequencies in Mongolia.

HAPLOGROUP *M1*
Ancestral line: "Eve" → *L1/L0* → *L2* → *L3* → *M* → *M1*

When individuals carrying the *M* mutations left Africa, they crossed through the Arabian Peninsula and continued on to the Indian subcontinent and East Asia. *M1* did not continue east. They stopped in Arabia and turned back toward Africa. *M1* is distinguished by a unique series of mutations and a unique geographic distribution. In fact, *M1* individuals are characterized by a set of four unique mutations. *M1* has subsequently split into four subgroups within eastern Africa, again characterized by genetic differences between their members. While the overall *M1* group has an age estimate of around 60,000 years old, the four *M1* subgroups have all diverged within the last 10,000 to 20,000 years in eastern Africa.

Today, around 20 percent of eastern African mitochondrial lineages belong to haplogroup *M1,* and its distribution spans the Red Sea. It constitutes most of the Mediterranean haplogroup *M* individuals, and around 7 percent of all lineages of the Nile Valley.

Members of this group are largely absent among Indian and East Asian populations. Interestingly, both the Indian and East Asian haplogroup *M* and the East African *M1* lineages have similar ages. Therefore it is possible that haplogroup *M* individuals did in fact live in East Africa prior to their exodus, only to be replaced by subsequent populations there. Conversely, *M1* individuals could have arisen after the exodus from Africa, only to return across the Red Sea as other *M*-bearing individuals were heading east. The specific origin of both lineages remains a topic of great scientific interest.

HAPLOGROUP C
Ancestral line: "Eve" → *L1/L0* → *L2* → *L3* → *M* → *C*

One group of these early *M* individuals broke away in the Central Asian steppes and set out on their own journey following herds of game across vast spans of inhospitable geography. Around 50,000 years ago, the first members of haplogroup *C* began moving north into Siberia, the beginnings of a journey that would not stop until finally reaching both continents of the Americas.

The haplogroup arose on the high plains of Central Asia between the Caspian Sea and Lake Baikal. It is considered a characteristic Siberian lineage, and today accounts for more than 20 percent of the entire mitochondrial gene pool found there. Because of its old age and high frequency throughout northern Eurasia, it is widely accepted that this lineage was carried by the first humans to settle these remote areas.

Radiating out from the Siberian homeland, haplogroup *C*-bearing individuals began migrating into the surrounding areas and quickly headed south, making their way into northern and Central Asia. Haplogroup *C*'s frequency slowly declines the farther from Siberia one looks: It now comprises around 5 to 10 percent of the people in central Asia, and around 3 percent of the people living in East Asia.

Heading west out of Siberia, however, this gradual reduction in frequency comes to an abrupt end around the Ural Mountains and Volga River. To the west of the Urals, this haplogroup is observed at frequencies less than one percent, both in eastern and western Europe. It appears that haplogroup *C* individuals were quite unsuccessful at making it across these mountains, a strong example of the impact geographic barriers have on human migration, and thus on gene flow.

Owing to their experience in the harsh Siberian climate, haplogroup *C* individuals would have been ideally suited for the arduous crossing of the Bering land bridge during the last ice age, 15,000 to 20,000 years ago. Colder temperatures and a drier global climate locked much of the world's fresh water at the polar ice caps, making living conditions near impossible for much of the Northern Hemisphere. But an important result of this glaciation was that eastern Siberia and northwestern Alaska became temporarily connected. This group, fishing along the coastline, followed it.

Today, haplogroup *C* is one of five mitochondrial lineages found in aboriginal Americans in both North and South America. While the age of *C* is very old (around 50,000 years), the reduced genetic diversity found in the Americas indicates that those lineages arrived only within the last 15,000 to 20,000 years and quickly spread once there. Better understanding exactly how many waves of humans crossed into the Americas, and where they migrated to once there, remains the focus of much interest, and is central to the Genographic Project's ongoing research in the Americas.

HAPLOGROUP *D*

Ancestral line: "Eve" → *L1/L0* → *L2* → *L3* → *M* → *D*

Around 50,000 years ago, another group of *M* individuals broke away in the Central Asian steppes to move into East Asia. The first members of haplogroup *D* began moving eastward and their descendants ultimately reached both North and South America.

Like haplogroup *C*, this haplogroup also lived on the high plains of central Asia between the Caspian Sea and Lake Baikal. *D* is considered a characteristic east Eurasian lineage, and today is the predominant haplogroup in East Asia, accounting for around 20 percent of the entire mitochondrial gene pool there.

Fanning out from the Central Asian homeland, haplogroup *D*-bearing individuals began migrating into the surrounding areas and quickly headed south, making their way throughout East Asia. In northern Asia *D* is present in more than 20 percent of the population, and is found at around 17 percent in Southeast Asia. Because of its old age and high frequency throughout east Eurasia, it is widely accepted that this lineage was carried by the first humans to settle the region.

Haplogroup *D* has a gradual reduction in frequency moving west across Eurasia. It is found in 10 to 20 percent of central Asians, with several of the lineages there matching exactly with sequences found in the East, representing a much more recent mixture. These lineages were likely introduced within the last 5,000 years, possibly during the ancient Silk Road trading thoroughfare that connected the entire Eurasian continent.

Haplogroup *D* was also one of those lineages that crossed into North America and became one of five mitochondrial lineages found in Native Americans of North and South America.

HAPLOGROUP *Z*
Ancestral line: "Eve" → *L1/L0* → *L2* → *L3* → *M* → *Z*

About 30,000 years ago, the first members of haplogroup *Z* began moving north into Siberia, the beginnings of a journey into much of eastern Asia. A characteristic Siberian lineage, haplogroup *Z* also inhabited the high plains of central Asia between the Caspian Sea and Lake Baikal. Today it accounts for around 3 percent of the entire mitochondrial gene pool found there. Because of its old age and frequency throughout northern Eurasia, it is widely accepted that this lineage was carried by the first humans to settle these remote areas.

From their Siberian homeland, haplogroup *Z*-bearers began migrating into the surrounding areas and headed south into northern and central Asia. Today haplogroup *Z* comprises around

2 percent of the East Asian population. But west of Siberia, near the Ural Mountains and Volga River, the frequency of haplogroup Z drops off drastically (it's observed at frequencies less than 1 to 2 percent). Z's attempts to establish themselves on the other side were largely unsuccessful.

Siberian mitochondrial lineages, such as C and D, with similar geographic distributions to Z made it successfully into the Americas. However, Z individuals did not, or at least their descendants have not survived to the present day. If in fact Z-bearing individuals were among the first humans to brave the Bering Strait, the number of people carrying that lineage was so low that their contribution to the gene pool in the Americas has been lost.

HAPLOGROUP N
Ancestral line: "Eve" → L1/L0 → L2 → L3 → N

Haplogroup N, like M, is one of two groups that descend directly from haplogroup L3. The first of these groups, M, made up the first great wave of human migration to leave Africa. The second great wave, also of L3 individuals, moved north rather than east and left the African continent across the Sinai Peninsula. Faced with the harsh desert conditions of the Sahara, these people likely followed the Nile basin, which would have proved a reliable water and food supply in spite of the surrounding desert and its frequent sandstorms. Descendants of these migrants eventually formed haplogroup N.

Early members of this group lived in the eastern Mediterranean region and western Asia, where they likely coexisted for a time with other hominids such as Neandertals. Excavations in Israel's Kebara Cave (Mount Carmel) have unearthed Neandertal skeletons as recent as 60,000 years old, indicating that there was both geographic and temporal overlap of these two hominids.

Some members bearing mutations specific to haplogroup *N* formed many groups of their own which went on to populate much of the rest of the globe. These descendants are found throughout Asia, Europe, India, and the Americas. However, because almost all of the mitochondrial lineages found in the Near East and Europe descend from *N*, it is considered a western Eurasian haplogroup.

After several thousand years in the Near East, members of this group began moving into unexplored nearby territories, following large migrating game herds across vast plains. These groups broke into several directions and made their way into territories surrounding the Near East. Today, descendants of haplogroup *N* individuals who headed west are prevalent in Turkey and the eastern Mediterranean; they are found farther east in parts of central Asia and the Indus Valley of Pakistan and India. And members of the haplogroup who headed north out of the Levant across the Caucasus Mountains have remained in southeastern Europe and the Balkans. Importantly, descendants of these people eventually went on to populate the rest of Europe, and today comprise the most frequent mitochondrial lineages found there.

HAPLOGROUP *N1*
Ancestral line: "Eve" → *L1/L0* → *L2* → *L3* → *N* → *N1*

In addition to a wide geographic distribution similar to *N*, this haplogroup is significant because its members constitute one of the four major Ashkenazi Jewish founder lineages. "Ashkenazi" refers to Jews of mainly central and eastern European ancestry. Most historical records indicate that the founding of this population took place in the Rhine Basin and subsequently underwent vast population expansions. The Ashkenazi population was estimated at approximately 25,000 individuals around A.D. 1300, whereas that number had increased to about 8,500,000 individuals by the turn of the 20th century.

Around half of all Ashkenazi Jews trace their mitochondrial lineage back to one of four women, and haplogroup *N1* represents one. It is seldom found in non-Ashkenazi populations, although it does appear at frequencies of roughly 3 percent or higher in those from the Levant, Arabia, and Egypt, which indicates a strong genetic role in the Ashkenazi founder event. Today, haplogroup *N1* is the second most common in Ashkenazi Jews and is shared by around 800,000 people.

HAPLOGROUP *A*
Ancestral line: "Eve" → *L1/L0* → *L2* → *L3* → *N* → *A*

Around 50,000 years ago, the first members of haplogroup *A* began moving east across Siberia, the beginnings of a journey that did not stop until finally reaching both continents of the Americas.

Haplogroup *A* likely arose on the high plains of central Asia. As groups moved east, they carried haplogroup *A* with them and it was spread to several different areas in East Asia. Haplogroup *A* was first found among aboriginal American populations and has played an important role in allowing geneticists to use DNA mutations to date prehistoric migrations. With few exceptions, haplogroup *A* is the only lineage carried by Eskimos, an indigenous group native to Siberia, Alaska, and Canada. Because a reliable date of around 11,000 years ago has been assigned to their colonization of these territories, DNA mutations specific to the Eskimo have allowed geneticists to infer a molecular clock for human mutations. It has also enabled geneticists to place accurate estimates not only on the Eskimos and the peopling of the Americas, but on prehistoric human migrations throughout the world.

HAPLOGROUP B
Ancestral line: "Eve" → *L1/L0* → *L2* → *L3* → *N* → *B*

Around 50,000 years ago, the first members of haplogroup *B* began moving into East Asia and ultimately reached both continents of the Americas and much of Polynesia. This haplogroup likely arose on the high plains of central Asia between the Caspian Sea and Lake Baikal. It is one of the founding East Asian lineages and, along with haplogroups *F* and *M,* comprises around three-quarters of all mitochondrial lineages found there today.

Radiating out from the central Asian homeland, haplogroup *B*-bearing individuals began migrating into the surrounding areas and quickly headed south, making their way throughout East Asia. Today haplogroup *B* is found in around 17 percent of people from Southeast Asia, and around 20 percent of the entire Chinese gene pool. It exhibits a very wide distribution along the Pacific coast, from Vietnam to Japan, as well as at lower frequencies (about 3 percent) among native Siberians. Because of its old age and high frequency throughout east Eurasia, it is widely accepted that this lineage was carried by the first humans to settle the region. Today, haplogroup *B* is also one of five mitochondrial lineages found in Native Americans, both North and South America.

Recent population expansions appear to have brought a subgroup of haplogroup *B* lineages from Southeast Asia into Polynesia. This lineage is referred to as *B4* and is characterized by a set of mutations that took a significant amount of time to accumulate on the Eurasian continent. This closely related subset of lineages likely spread from Southeast Asia into Polynesia within the last 5,000 years, and is seen extensively throughout the islands at high frequency. Intermediate lineages—those containing some, but not all, of the *B4* mutations—are found in Vietnamese, Malaysian, and Bornean populations, further supporting the likelihood that the Polynesian lineages originated in these parts of Southeast Asia.

HAPLOGROUP I

Ancestral line: "Eve" → $L1/L0$ → $L2$ → $L3$ → N → I

Haplogroup I descended from the N haplogroup, whose descendants live in high frequencies in northern Europe and northern Eurasia. I individuals used the Near East as a "home base" of sorts, radiating from that region to populate much of the rest of the world. Today, members in the Near East belonging to haplogroup I have more divergent lineages than those found in northern Europe, indicating a greater time in the Near East for those lineages to accumulate mutations. Therefore, early members of haplogroup I likely moved north across the Caucasus, their lineages being carried into Europe for the first time during the middle Upper Paleolithic.

This wave of migration into western Europe marked the appearance and spread of what archaeologists call Aurignacian culture, which is distinguished by significant innovations in tool manufacture and invention; people began using a broader set of tool types, such as end-scrapers for preparing animal skins and tools for woodworking. In addition to stone, these modern humans used bone, ivory, antler, and shells to make tools. Jewelry, often an indicator of status, appears in Aurignacian culture as well; bracelets and pendants made of shells, teeth, ivory, and carved bone suggest the beginnings of a more complex social organization.

Today, only about 10 percent of the mitochondrial lineages found in Europe reflect the original early Upper Paleolithic movements into the continent, and about 20 percent reflect the more recent Neolithic movements. The rest of the European mtDNA, including the I lineage, are the result of migrations into Europe during the middle Upper Paleolithic around 25,000 years ago. These took part in the postglacial expansions around 15,000 years ago as the ice sheets receded during the late Upper Paleolithic.

HAPLOGROUP W

Ancestral line: "Eve" → *L1/L0* → *L2* → *L3* → *N* → *W*

Members of haplogroup *W* also descended from haplogroup *N* and migrated into Europe from the Near East. Like haplogroup *I* descendants, *W* members who live in the Near East today have more divergent lineages than those found in northern Europe, indicating a longer habitation in the Near East for those lineages to accumulate more mutations. Early members of haplogroup *W* likely moved north across the Caucasus during the middle Upper Paleolithic. Like haplogroup *I,* Aurignacian culture is also associated with *W* members.

HAPLOGROUP X

Ancestral line: "Eve" → *L1/L0* → *L2* → *L3* → *N* → *X*

Haplogroup *X* is primarily composed of two distinct subgroups, *X1* and *X2,* whose widespread and sporadic distribution have been the cause of much debate. *X1* is largely found in North and East Africa. The other subgroup, *X2,* is spread widely throughout western Eurasia. *X2* makes up around 2 percent of the European mtDNA lineages and is more strongly present in the Near East, Caucasus region, and Mediterranean Europe. In some western Eurasian groups, it is found at significant frequencies of 10 to 25 percent, though this is likely due to expansion events following the last ice age around 15,000 years ago.

The real controversy surrounding haplogroup *X* is its place as one of the five haplogroups found in the indigenous peoples of the Americas, where it is found exclusively in North America at varying frequencies. In the Ojibwa from the Great Lakes region it is found at around 25 percent, in the Sioux at around 15 percent, the Nuu-Chah-Nulth at more than 10 percent, and in the Navajo at 7 percent. But *X2* is almost entirely absent from Siberia, the proposed land route of the first migrations into the New World.

Unlike the four main Native American haplogroups (*A, B, C,* and *D*), haplogroup *X* is entirely absent in East Asian populations, indicating that it played no role in colonizing of those regions. However, its age estimate in the Americas, around 15,000 years old, does indicate that members of this group were among the first modern humans there. One group in southern Siberia has been found containing *X* lineages, but the almost identical sequences in these individuals indicate that these *X*-bearing individuals are the result of recent gene flow into the area during the more recent Neolithic expansions of agriculturalists.

The widespread geographic distribution of this haplogroup, and its virtual absence in Siberia despite a prevalence among some Native American groups, promises to remain the focus of much scientific interest as anthropologists look to re-create the migrations that first brought humans to all corners of the globe.

HAPLOGROUP *R*
Ancestral line: "Eve" → *L1/L0* → *L2* → *L3* → *N* → *R*

The *R* clan is a group of individuals who descend from a woman in the *N* branch of the tree. This woman was the common ancestor of what can be described as a western Eurasian lineage, the descendants of whom live in high frequencies in the Anatolian/Caucasus and Iranian regions.

Haplogroup *R*'s history is complicated, however, because it is found almost everywhere, and its origins are quite ancient. In fact, the ancestor of haplogroup R lived relatively soon after humans moved out of Africa during the second wave, and her descendants undertook many of the same migrations as her own group, N.

Because the two groups lived side by side for thousands of years, it is likely that the migrations radiating out from the Near East comprised individuals from both of these groups. They simply moved together, bringing their N and R lineages to the same places around the same time. Their genetic lines became quickly entangled, and geneticists are currently working to unravel the different stories of haplogroups N and R, since they are found in many of the same far-reaching places.

But haplogroup R did give rise to many distinct groups that are common today. While some groups of R individuals moved across the Middle East into central Asia and the Indus Valley, others moved south, heading back into the African homeland from where their ancestors had recently departed. R members also moved north across the rugged Caucasus Mountains, their lineages being carried into Europe for the first time by the Cro-Magnon. Their arrival in Europe around 35,000 years ago heralded the end of the era of the Neandertals, a hominid species that inhabited Europe and parts of western Asia from about 230,000 to 29,000 years ago. Better communication skills, weapons, and resourcefulness probably enabled them to outcompete Neandertals for scarce resources. Today descendants of haplogroup R dominate the European mitochondrial genetic landscape, owing to more than 75 percent of the lineages found there.

HAPLOGROUP F
Ancestral line: "Eve" → $L1/L0$ → $L2$ → $L3$ → N → R → F

Haplogroup F likely developed on the high plains of central Asia between the Caspian Sea and Lake Baikal. It is one of the founding East Asian lineages and, along with haplogroups B and M, comprises around three-quarters of all mitochondrial lineages found there today.

Around 50,000 years ago, the first members of haplogroup *F* began moving into East Asia after they broke away from their *R* ancestors in Central and Southeast Asia. They would eventually spread out and reach widespread distribution throughout Southeast Asia, where today they make up more than 25 percent of the Southeast Asians.

Today haplogroup *F* shows its greatest sequence diversity in Vietnam. It exhibits a very wide distribution along the Pacific coast: It appears in Filipino and aboriginal Taiwanese populations, and while having decreased frequency distributions farther from Southeast Asia, it is found as far north as the Evenk of central Siberia and as far south as the Kadazan of Borneo.

Haplogroup *F* occurs in some coastal populations of Papua New Guinea, and much interest remains about the original spread of this lineage into Indonesia and perhaps Polynesia. Because *F* exhibits lower sequence diversity in Melanesia and Polynesia it is unlikely that this lineage was carried by the native colonizers that brought Austronesian speakers to the South Pacific islands. Instead, it is possible that carriers spread across Southeast Asia during the expansion of Sino-Tibetan languages, which took place sometime around 6,000 to 8,000 years ago. Better understanding the human prehistory of Southeast Asia, including Indonesia, Melanesia, and Polynesia, remains of keen interest to anthropologists, and is a primary focus of the Genographic Project's ongoing research in those areas.

HAPLOGROUP *PRE-HV*
Ancestral line: "Eve" → *L1/L0* → *L2* → *L3* → *N* → *R* → *pre-HV*

Individuals in haplogroup *pre-HV* can be found all around the Red Sea and widely throughout the Near East. While this genetic lineage is common in Ethiopia and Somalia, individuals from this group are found at highest frequency in Arabia. Because of their close genetic and geographic proximity to other western Eurasian clusters, members of this group living in eastern Africa are the likely result of more recent migrations back into the continent.

Haplogroup *pre-HV* descends from *R* and is sometimes referred to as *R0*. These descendants live in high frequencies in the Anatolian/Caucasus region and Iran. While members of this group can also be found in the Indus Valley near the Pakistan-India border, their presence is considered the result of a subsequent migration eastward of individuals out of the Near East.

As we've seen, descendants from haplogroups *N* and *R* used the Near East as a home base of sorts, radiating from that region to populate much of the rest of the world. Their descendants comprise all of the western Eurasian genetic lineages, and about half of the eastern Eurasian mtDNA gene pool. Some individuals moved across the Middle East into central Asia and the Indus Valley near western India. Some moved south, heading back into the African homeland from where their ancestors had recently departed.

Other members of *pre-HV* moved north across the Caucasus Mountains and west across Anatolia, their lineages being carried into Europe for the first time by the Cro-Magnon.

Over the course of several thousand years, descendants of *pre-HV* began to split off and form their own groups. Of primary importance is a group called *HV*, which gave rise to the two most prevalent female lineages found in western Europe, *H* and *V* (see below). While all of these lineages existed around 20,000 years ago, they did not dominate the female genetic landscape until after the climate window closed again. Expanding ice sheets forced people to move south to Spain, Italy, and the Balkans, and it was only when the ice retreated and temperatures became warmer, beginning about 12,000 years ago, that members of these groups moved north again and lived in the places that had become inhospitable during the previous ice age.

HAPLOGROUP HV

Ancestral line: "Eve" → $L1/L0$ → $L2$ → $L3$ → N → R → pre-HV → HV

While some descendants of these ancestral lineages moved out across central Asia, the Indus Valley, and even back into Africa, HV ancestors remained in the Near East. Descending from haplogroup pre-HV, they formed a new group, characterized by a unique set of mutations.

Haplogroup HV is a west Eurasian haplogroup found throughout the Near East, including present-day Turkey and the Caucasus Mountains of southern Russia and the republic of Georgia. It is also found in parts of East Africa, particularly in Ethiopia, where its presence there indicates recent Near Eastern gene flow, likely the result of the Arab slave trade over the last two millennia.

Much earlier, around 30,000 years ago, some members of HV moved north across the Caucasus Mountains and west across Turkey to carry their lineages into Europe.

Around 15,000 to 20,000 years ago, during the last ice age, early Europeans retreated to the warmer climates of the Iberian Peninsula, Italy, and the Balkans, where they waited out the cold spell. Their population sizes were drastically reduced, and much of the genetic diversity that had previously existed in Europe was lost. After the ice sheets began to retreat, humans moved north again and recolonized western Europe. The two most frequent mitochondrial lineages carried by these expanding groups were H and V. Today they dominate the western European mitochondrial landscape, making up almost 75 percent of all European lineages.

HAPLOGROUP *HV1*

Ancestral line: "Eve" → *L1/L0* → *L2* → *L3* → *N* → *R* → *pre-HV* → *HV1*

Descending from haplogroup *pre-HV,* haplogroup *HV1* formed a new group around 30,000 years ago. Like *HV, HV1* can be found at its highest frequencies throughout the Near East, including Anatolia (present-day Turkey) and the Caucasus Mountains of southern Russia and the republic of Georgia. Some members of this haplogroup crossed the rugged Caucasus Mountains in southern Russia, moved on to the steppes of the Black Sea, and then westward into regions that comprise the present-day Baltic states and western Eurasia. This grassland then served as the home base for subsequent movements north and west. Today, members of these lineages are found in eastern Europe and the eastern Mediterranean region.

Despite geographic proximity to the Iberian Peninsula, members of this group did not give rise to the two dominant western European lineages, *H* and *V.* This indicates that while many early Europeans were struggling to survive through the last ice age, *HV1* ancestors waited it out safely to the temperate south in the familiar Near East. Interestingly, *HV1* is also found in parts of East Africa, particularly in Ethiopia, where its presence there indicates recent Near Eastern gene flow, likely the result of the Arab slave trade over the last two millennia.

HAPLOGROUP *H*

Ancestral line: "Eve" → *L1/L0* → *L2* → *L3* → *N* → *R* → *pre-HV* → *HV* → *H*

As humans began to repopulate western Europe after the ice age, by far the most frequent mitochondrial lineage carried by these expanding groups was haplogroup *H,* which came to dominate the European female landscape.

Today haplogroup *H* comprises 40 to 60 percent of the gene pool of most European populations. In Rome and Athens, for example, *H* is found in about 40 percent of the entire population, and it exhibits similar frequencies throughout western Europe. Moving eastward the frequencies of *H* gradually decrease, illustrating the migratory path these settlers followed as they left the Iberian Peninsula after the ice sheets had receded. Haplogroup *H* is found at around 25 percent in Turkey and around 20 percent in the Caucasus Mountains.

While haplogroup *H* is considered the western European lineage due to its high frequency there, it is also found much farther east. Today it comprises around 20 percent of southwest Asian lineages, about 15 percent of people living in central Asia, and around 5 percent in northern Asia.

Importantly, the age of haplogroup *H* lineages differs quite substantially between those seen in the West compared with those found in the East. In Europe its age is estimated at 10,000 to 15,000 years old, and while *H* made it into Europe substantially earlier (30,000 years ago), reduced population sizes resulting from the ice age significantly reduced its diversity there, and thus its estimated age. In Central and East Asia, however, its age is estimated at around 30,000 years old, meaning the lineage made it into those areas during some of the earlier migrations out of the Near East.

HAPLOGROUP *V*
Ancestral line: "Eve" → *L1/L0* → *L2* → *L3* → *N* → *R* → *pre-HV* → *HV* → *V*

Today, haplogroup *V* tends to be restricted to western, central, and northern Europe. Its age is estimated at around 15,000 years old, indicating that it likely arose during the 5,000 years or so that humans were confined to the European refugia during the last ice age. It is found in around 12 percent of Basques, an isolated population in northern Spain, and around 5 percent in many other western European populations. It is also found in Algeria and Morocco, indicating that these humans migrating out of the Iberian Peninsula also headed south across the Strait of Gibraltar and into North Africa. Its genetic diversity reduces gradually moving west to east, indicating the migratory direction these groups followed during the recolonization.

Interestingly, haplogroup V attains its highest frequency in the Skolt Saami of northern Scandinavia, a group of hunter-gatherers who follow the reindeer herds seasonally from Siberia to Scandinavia and back. While V makes up about half the mitochondrial lineages in the Saami, its genetic diversity is considerably reduced compared to that observed in western Europe, and was likely introduced into the Saami within the past several thousand years.

HAPLOGROUP J

Ancestral line: "Eve" → $L1/L0$ → $L2$ → $L3$ → N → R → J

This group of individuals also descended from a woman in the R branch of the tree. The divergent genetic lineage that constitutes haplogroup J indicates that she lived sometime around 40,000 years ago. Haplogroup J has a very wide distribution, and is present as far east as the Indus Valley bordering India and Pakistan, and as far south as the Arabian Peninsula. It is also common in eastern and northern Europe. Although this haplogroup was present during the early and middle Upper Paleolithic, J is largely considered one of the main genetic signatures of the Neolithic expansions.

While groups of hunter-gatherers and subsistence fishermen had been occupying much of Eurasia for tens of thousands of years, around 10,000 years ago a group of modern humans living in the

Fertile Crescent (present-day eastern Turkey and northern Syria) began domesticating the plants, nuts, and seeds they had been collecting. What resulted were the world's first agriculturalists, and this new cultural era is typically referred to as the Neolithic.

Groups of individuals able to support larger populations with this reliable food source began migrating out of the Middle East, bringing their new technology with them. By then, humans had already settled much of the surrounding areas, but this new agricultural technology proved too successful to ignore, and the surrounding groups quickly copied these new immigrants. Agriculture was quickly and widely adopted, but the lineages carried by these Neolithic expansions are found today at low frequencies.

Haplogroup *J* has greater diversity in the Near East than in Europe, indicating a homeland for *J*'s most recent common ancestor around the Levant, a coastal region in Lebanon. It reaches its highest frequency in Arabia, comprising around 25 percent of the Bedouin and Yemeni. But genetic evidence indicates that the higher incidence is more reflective of low population sizes or the occurrence of a founder event than this region actually being the geographic origin of haplogroup *J*.

HAPLOGROUP *K*
Ancestral line: "Eve" → *L1/L0* → *L2* → *L3* → *N* → *R* → *K*

K individuals also descend from a woman in the *R* branch of the tree. Because of the great genetic diversity found in haplogroup *K*, it is likely that she lived around 50,000 years ago. Interestingly, her descendants gave rise to several different subgroups, some of which exhibit very specific geographic homelands. The very old age of these subgroups has led to a wide distribution; today they harbor specific European, northern African, and Indian components, and are found in Arabia, the northern Caucasus Mountains, and throughout the Near East.

While some members of this haplogroup headed north into Scandinavia, or south into North Africa, most members of haplogroup *K* stem from a group of individuals who moved northward out of the Near East. These women crossed the rugged Caucasus Mountains in southern Russia, and moved on to the steppes of the Black Sea.

Like the *N1* lineage, haplogroup *K* is very significant because it and its subgroups also constitute three of the four major Ashkenazi Jewish founder lineages. Around half of all Ashkenazi Jews trace their mitochondrial lineage back to one of four women, and haplogroup *K* represents a lineage that gave rise to three of them. While this lineage is found at a smaller frequency in non-Ashkenazi Jews, the three *K* lineages that helped found the Ashkenazi population are seldom found in other populations. While virtually absent in Europeans, they appear at frequencies of 3 percent or higher in groups from the Levant, Arabia, and Egypt. This indicates a strong genetic role in the Ashkenazi founder event, which likely occurred in the Near East. Today, *K* has given rise to three of the four most common haplogroups in Ashkenazi Jews and is currently shared by more than three million people.

HAPLOGROUP *T*
Ancestral line: "Eve" → *L1/L0* → *L2* → *L3* → *N* → *R* → *T*

The divergent genetic lineage that constitutes haplogroup *T* indicates that the woman they descended from lived sometime around 40,000 years ago. Haplogroup *T* has a very wide distribution, and is present as far east as the Indus Valley and as far south as the Arabian Peninsula. It is also common in eastern and northern Europe. Although this haplogroup was present during the early and middle Upper Paleolithic, *T* is largely considered one of the main genetic signatures of the Neolithic expansions.

HAPLOGROUP *U*

Ancestral line: "Eve" → *L1/L0* → *L2* → *L3* → *N* → *R* → *U*

Because of the great genetic diversity found in haplogroup *U,* it is likely that it developed around 50,000 years ago. The very old age of haplogroup *U* and its subgroups has led to a wide distribution; today they harbor specific European, northern African, and Indian components, and are found in Arabia, the northern Caucasus Mountains, and throughout the Near East.

While some members headed north into Scandinavia, or south into North Africa, most haplogroup *U* individuals come from a group that moved northward out of the Near East. These women crossed the rugged Caucasus Mountains and moved on to the steppes of the Black Sea. From there, they continued west until they reached grasslands near the present-day Baltic states and western Eurasia. This region became their home base for further movements north and west. Today, haplogroup *U* members are found in Europe and the eastern Mediterranean at frequencies of almost 7 percent of the population.

HAPLOGROUP *U5*

Ancestral line: "Eve" → *L1/L0* → *L2* → *L3* → *N* → *R* → *U* → *U5*

The most recent common ancestor for all *U5* individuals broke off from the rest of the group and headed north into Scandinavia. Even though *U5* is descended from an ancestor in haplogroup *U,* it is also an ancient lineage, estimated to be around 50,000 years old. *U5* is quite restricted in its variation to Scandinavia, particularly to Finland. This is likely the result of the significant geographical, linguistic, and cultural isolation of the Finnish populations, which would have limited geographic distribution of this subgroup and kept it fairly isolated. The Saami, reindeer hunters who follow the herds from

Siberia to Scandinavia each season, have the *U5* lineage at a very high frequency (around 50 percent) indicating that it may have been introduced during their movements into these northern territories.

The *U5* lineage is found outside of Scandinavia, though at much lower frequencies and at lower genetic diversity. Interestingly, the *U5* lineage found in the Saami has also been found in some North African Berber populations in Morocco, Senegal, and Algeria. Finding similar genetic lineages in populations living thousands of miles apart is certainly unexpected, and is likely the result of movements that occurred 15,000 years ago when the last ice age came to an end.

In addition to being present in some parts of North Africa, *U5* individuals also live sporadically in the Near East at 2 percent—about one-fifth as frequent as in parts of Europe—and are completely absent from Arabia. Their distribution in the Near East is largely confined to surrounding populations, such as Turks, Kurds, Armenians, and Egyptians. Because these individuals contain lineages that first evolved in Europe, their presence in the Near East is the result of a back-migration of people who left northern Europe and headed south, as though retracing the migratory paths of their own ancestors.

HAPLOGROUP *U6*
Ancestral line: "Eve" → *L1/L0* → *L2* → *L3* → *N* → *R* → *U* → *U6*

Haplogroup *U6* is estimated to be around 50,000 years old. *U6* individuals, who broke off from haplogroup *U* while still in the Near East, are found predominantly in North Africa. While other members from haplogroup *U* were heading northward into Europe and Scandinavia, these people headed west along the southern Mediterranean coast. Today, they are found primarily in northern Africa and constitute around 10 percent of the people living there.

The age estimate for this group is consistent with the physical-anthropological view that the original inhabitants of North Africa were closely related to the Cro-Magnon settlers who first colonized Europe during the Upper Paleolithic. These people constructed huts to withstand the harsh climates and dry conditions. They used relatively advanced tools of stone, bone, and ivory. Jewelry, carvings, and intricate, colorful rock paintings bear witness to their surprisingly advanced culture.

After the last ice age, *U6* ancestors moved across the Strait of Gibraltar for the first time, which allowed for some gene flow between North Africa and the western Mediterranean. This bidirectional movement resulted in some *U6* lineages being found in parts of western Europe (in southern Spain and France) despite these being predominantly North African lineages.

Y-CHROMOSOME HAPLOGROUPS

HAPLOGROUP *A*
Ancestral line: "Adam" → *M91*

The most diverse of all Y-chromosome lineages, haplogroup *A* dates back to roughly 60,000 years ago and is defined by the marker *M91*. Genetic diversity increases with age, so *M91* provides a genetic link to the earliest common male ancestor of all humans, "Adam."

Today, many individuals carrying the marker *M91* live in Ethiopia, the Sudan, and southern regions in Africa. They often practice cultural traditions that are representative of the ways of life of their distant ancestors. For example, some live in traditional hunter-gatherer societies once common to all humans. They also may still speak ancient click languages, like those of the San Bushmen of the Kalahari and the Hadzabe of Tanzania.

Other members of this haplogroup are the relatively recent descendants of these people. Within the last 2,000 to 3,000 years, these ancient cultures have been greatly reduced in size by the impressive growth of Africa's Bantu culture.

HAPLOGROUP B
Ancestral line: "Adam" → *M60*

M60 defines haplogroup *B*, an ancient African lineage that originated some 50,000 to 60,000 years ago. As with most very old lines of descent, it has a broad dispersal, is found today across the African continent, and is shared by many different African peoples. Often these unique populations and cultures, like that of the Biaka and Mbuti Pygmies, are themselves quite ancient.

HAPLOGROUP C
Ancestral line: "Adam" → *M168* → *M130*

Approximately 50,000 years ago, probably in southern Asia, a man was born carrying marker *M130*. His recent ancestors had just begun the first major wave of migration out of Africa. They followed the African coastline through the southern Arabian Peninsula, India, Sri Lanka, and Southeast Asia. Some members of the group would ultimately cross the Torres Strait to populate distant Australia. Because these early travelers were already quite adept at exploiting coastal resources, they didn't need to learn new skills as they followed the coastline, and the migration to Australia occurred quickly, taking less than 5,000 years.

Not everyone in the clan made the full journey to Australia. Many remained on the Southeast Asian coastline and gradually moved inland, migrating northward over thousands of years where their descendants can still be found in eastern Asia, particularly in Mongolia and Siberia.

Sometime in the last 10,000 years, a group of these descendants living in northern China or southeastern Siberia climbed into their boats and migrated along the coast of the Pacific Rim to reach North America. Evidence that supports this migration theory is provided by the distribution of the Na-Dene languages, such as Tlingit and Navajo, which are limited to the western half of North America. In Na-Dene populations, particularly in western Canada and southwest United States, as many as 25 percent of the men carry marker *M130*.

HAPLOGROUP C3
Ancestral line: "Adam" → *M168* → *M130* → *M217*

The genetic marker *M217* arose in people who lived among ancient East Asian populations about 20,000 years ago. From East Asia the descendants of *M217* carried the marker west and south toward central Asia.

Genealogists believe that this lineage spread, at least in part, through the legendary Mongol conquests of Genghis Khan during the 12th and 13th centuries A.D. Around 16 million men (about one in every ten, or 0.5 percent of the world's male population) living in Central and East Asia descend from this line and presumably from this shared ancestor.

YAP: AN ANCIENT MUTATION
Ancestral line: "Adam" → *M168* → *YAP*

Y *Alu* Polymorphism, or *YAP* for short, is characterized by a mutational event known as an *Alu* insertion, a 300-nucleotide fragment of DNA that on rare occasion gets inserted into different parts of the human genome during cell replication. A man living around 50,000 years ago acquired this fragment on his Y chromosome and passed it on to his descendants.

YAP can be found in populations that occurred around northeast Africa and is the most common of the three ancient genetic branches found in sub-Saharan Africa. Over time the *YAP* lineage split into two distinct groups: One, haplogroup *D*, is found in Asia and is defined by the *M174* mutation. The other, haplogroup *E*, is found primarily in Africa and the Mediterranean and is defined by marker *M96*.

HAPLOGROUP *D*
Ancestral line: "Adam" → *M168* → *YAP* → *M174*

Ancestors of haplogroup *D* may have accompanied haplogroup *C* on the first major wave of migration out of Africa. But some experts say that they may have made the trek at a later time. Pockets of haplogroup *D* still live along the ancient route, particularly in Southeast Asia and the Andaman Islands, but not in India.

Population distribution also suggests that there were two later migrations after haplogroup *D* originally left Africa. The first carried ancestors north along the East Asian coast into Japan. A more recent migration occurred within the last several thousand years and brought descendants of this lineage into Tibet from Mongolia.

HAPLOGROUP *D1*
Ancestral line: "Adam" → *M168* → *YAP* → *M174* → *M15*

M15, the genetic marker that defines haplogroup *D1*, first appeared in humans some 30,000 years ago, probably in Southeast Asia. Descendants of the lineage's patriarch moved northwest toward Tibet within the last several thousand years. Tibet is home to the highest frequency of haplogroup *D1*, though the lineage is still found in Southeast Asia as well.

HAPLOGROUP *D2*
Ancestral line: "Adam" → *M168* → *YAP* → *M174* → *P37.1*

About 30,000 years ago in Southeast Asia, the genetic marker *P37.1* first appeared. Today that marker defines the haplogroup *D2,* and carriers are identified as descendants of the *M174* migrants who undertook a gradual northward migration and eventually reached Japan. Today haplogroup *D2* is most common there, occurring at frequencies of 50 percent in some Japanese populations.

HAPLOGROUP *E*
Ancestral line: "Adam" → *M168* → *YAP* → *M96*

The marker *M96* first appeared in northeast Africa 30,000 to 40,000 years ago. Its precise origins are still unclear. What is known is that the during the second wave of migration out of Africa, about 50,000 years ago, a group of travelers called the Middle Eastern Clan—largely descended from a man born with marker M89 (see "Haplogroup F")—left the continent. They headed north and eventually settled in the Middle East. Members of haplogroup *E* may have accompanied the Middle Eastern Clan on this initial trip, but they also may have undertaken their own migration at a later date, following the same route previously traveled by the Middle Eastern Clan peoples.

HAPLOGROUP *E3A*
Ancestral line: "Adam" → *M168* → *YAP* → *M96* → *M2*

The man who gave rise to this lineage was born in Africa about 30,000 years ago. His descendants traveled south to sub-Saharan Africa. Experts hypothesize that this haplogroup dispersed eastward and southward from western Africa within the last 2,500 years, with the expansion of Bantu-speaking peoples from their homeland in West-Central Africa. Today it is found at frequencies of more than 70 percent in populations from Nigeria and Cameroon. *E3a* is also the most common lineage among African Americans.

HAPLOGROUP *E3B*
Ancestral line: "Adam" → *M168* → *YAP* → *M96* → *M35*

Around 20,000 years ago, the *M35* marker appeared in the Middle East among populations of the first farmers who helped spread agriculture from the Middle East into the Mediterranean region.

As the last ice age ended around 10,000 years ago, the climate became more conducive to plant production, helping spur the Neolithic Revolution. At this crucial moment, humans changed their way of living from nomadic hunter-gatherers to settled agriculturists. Early farming successes in the Fertile Crescent beginning around 8,000 years ago spawned population booms and encouraged migration throughout much of the Mediterranean world.

Control over their food supply marks a major turning point for the human species. Rather than small clans of 30 to 50 people who were highly mobile and informally organized, agriculture brought the first trappings of civilization. Occupying a single territory required more complex social organization, moving from the kinship ties of a small tribe to the more elaborate relations of a larger community. It spurred trade, writing, and calendars and pioneered the rise of modern, sedentary communities and cities.

HAPLOGROUP *F*
Ancestral line: "Adam" → *M168* → *M89*

The marker *M89* first appeared around 45,000 years ago in northeastern Africa or the Middle East, and today is found in more than 90 percent of all non-African men. The first people who left Africa likely followed a coastal route that eventually ended in Australia, but members of haplogroup *F,* however, appear to have followed the grasslands and plentiful game from eastern Africa to the Middle East and beyond. They were part of the second great wave of migration out of Africa.

While many of the descendants of *M89* remained in the Middle East, others continued to follow the great herds of antelope, mammoths, and other game through what is now modern-day Iran to the vast steppes of central Asia. These semi-arid grass-covered plains formed an ancient "superhighway" stretching from eastern France to Korea. These ancestors, having migrated north out of Africa into the Middle East, then traveled both east and west along this central Asian route. A smaller group continued moving north from the Middle East to Anatolia and the Balkans, trading familiar grasslands for forests and high country.

HAPLOGROUP *G*
Ancestral line: "Adam" → *M168* → *M89* → *M201*

Members of haplogroup G share a marker, *M201,* that arose around 30,000 years ago in a man born along the eastern edge of the Middle East, perhaps close to the Himalayan foothills in Pakistan or India. His descendants are few, and members of this clan are rarely found at frequencies greater than a few percent in these populations. Members of haplogroup *G* lived in the Indus Valley before the expansion of Neolithic farmers into the region. It was once thought that the advancing farmers displaced or even eliminated the hunter-gatherers of haplogroup *G.* However, DNA studies now show that, despite living at high altitudes and having low population densities, individuals from haplogroup *G* were able to survive the Neolithic expansion by learning to farm and adopting the Neolithic culture. Haplogroup *G* has three related "brother" haplogroups: *H, I,* and *J.* All of these probably arose between 20,000 and 30,000 years ago, and their spread was at least partially due to the expansion of agriculture.

HAPLOGROUP *G2*
Ancestral line: "Adam" → *M168* → *M89* → *M201* → *P15*

About 30,000 years ago, the genetic marker *P15* emerged and now defines the haplogroup *G2*. The *G2* lineage arose in the Middle East, though *P15* descendants soon spread westward through modern Turkey into southeastern Europe. The bulk of these migrations took place more than 15,000 years ago, before much of Europe was locked in ice during the last glacial maximum.

During the time when these ice sheets were largest, individuals living outside of the warmer refugia would have been unable to survive and thus effectively eliminated from the gene pool. This reduced the genetic diversity of the surviving populations, actually helping those lucky lineages to become fixed at higher frequencies in the subsequent generations.

When the glaciers finally began to recede, the *G2* lineage expanded northward and eastward to repopulate Europe. Evidence of these journeys can be seen by the marker's presence in western Eurasia.

HAPLOGROUP *H*
Ancestral line: "Adam" → *M168* → *M89* → *M69*

Ancestors of haplogroup *H* migrated along the steppe superhighway from the Middle East around 45,000 years ago, continuing toward India. On this journey, a boy was born approximately 30,000 years ago with the *M69* marker, which came to define this new lineage. Although *M69* is known as an "Indian Marker," this male ancestor may have been born in southern central Asia. His descendants were part of the first major inland settlement of India.

Geneticists believe that haplogroup *H* might have originated somewhere along the migration route of peoples carrying the

M20 Y-chromosome marker (see "Haplogroup L"). The peoples of these clans migrated along the steppe highway from the Middle East, and then moved south into India.

These were not the first humans to arrive in India, but they likely undertook the first major settlement of the region some 30,000 years ago. The earliest waves of African migrants had traveled along the Indian coastline 50,000 to 60,000 years before; some settled along the coastal route, but inland areas were largely populated by *H* members coming down from the north.

HAPLOGROUP *H1*
Ancestral line: "Adam" → *M168* → *M89* → *M69* → *M52*

The specific genetic marker that defines haplogroup *H1, M52*, is part of a largely Indian lineage. This marker made its first known appearance some 25,000 years ago in India. *M52* was part of the second major wave of human migration into India, long after a large wave of African migrants traveled along the Indian coastline 50,000 to 60,000 years ago.

Haplogroup *H1* ancestors appear to have arrived from the north on a journey from the Middle East and were likely the first to establish significant settlement in India. It became quickly established throughout the area and was successfully passed on to subsequent generations.

Today *H1* is found at frequencies as high as 25 percent in some Indian populations. Members of haplogroup *H1* are also found in lower frequencies in Iran and throughout much of southern central Asia.

HAPLOGROUP *I*
Ancestral line: "Adam" → *M168* → *M89* → *M170*

Ancestors of this group were part of the *M89* Middle Eastern clan that continued to migrate northwest into the Balkans and eventually spread into central Europe. These people may have been responsible for bringing the Gravettian culture to western Europe about 21,000 to 28,000 years ago.

Named after a site in La Gravette, France, Gravettian culture represented a new technological and artistic phase in western Europe. Archaeologists discovered sets of tools different from the preceding era (Aurignacian culture). These stone tools had a distinctive, small pointed blade, which humans used to hunt big game. Gravettian culture is also known for voluptuous carvings of big-bellied females often dubbed "Venus" figures. The small, frequently hand-sized sculptures appear to be of pregnant women and may have served as fertility icons, or emblems conferring protection of some sort, or may have represented goddesses. These early European ancestors used communal hunting techniques, created shell jewelry, and used mammoth bones to build their homes. Recent findings suggest that they may have discovered how to weave clothing using natural fibers as early as 25,000 years ago. Earlier estimates had placed weaving at about the same time as the emergence of agriculture, around 10,000 years ago.

The most recent common ancestor, the man who gave rise to marker *M170*, was born about 25,000 years ago. His descendants were later forced into the isolated refuge areas during the last blast of the ice age in the Balkans and Iberia. As the ice sheets covering much of Europe began to retreat, his descendants likely played a central role in recolonizing central and northern Europe.

HAPLOGROUP *I1A*
Ancestral line: "Adam" → *M168* → *M89* → *M170* → *M253*

Some 20,000 years ago this group, like many Europeans, sought refuge from the massive sheets of ice that covered much of the continent during the last ice age. They found temperate ice-free refugia on the Iberian Peninsula.

While this ancestral lineage was geographically isolated, the distinctive genetic marker *M253* appeared in one of its male members. As the Earth warmed and the glacial maximum passed, some 15,000 years ago the refugia's dwellers left the peninsula and began to repopulate other parts of Europe. They carried with them the unique genetic marker that defines haplogroup *I1a*.

Today this marker is still commonly found throughout northwest Europe. Because of its high frequency in western Scandinavia, it is likely that many Vikings descended from this line. The Viking raids on the British Isles might help to explain the dispersal of this lineage there as well.

HAPLOGROUP *I1B*

Ancestral line: "Adam" → *M168* → *M89* → *M170* → *P37.2*

Haplogroup *I1b* is further defined by a marker known as *P37.2*. This marker appeared in the Balkans about 15,000 years ago and is still most commonly found there today. The *P37.2* marker also likely distinguishes ancient human populations who migrated to Balkan refugia during the glacial maximum at the peak of the last ice age.

When the glaciers finally began to recede, the *I1b* lineage expanded northward and eastward into Europe, and carried the marker *P37.2* along for the ride. Evidence of these journeys can be seen by the marker's significant presence in central and eastern Europe. It is also possible that the Celtic expansions of the first millennium B.C. may have contributed to the spread of this lineage.

HAPLOGROUP *J*

Ancestral line: "Adam" → *M168* → *M89* → *M304*

The patriarch of haplogroup *J* was born around 15,000 years ago in the Fertile Crescent, a region today that includes Israel, the West Bank, Jordan, Lebanon, Syria, and Iraq. Today the *M304* marker appears at its highest frequencies in the Middle East, North Africa, and Ethiopia. In Europe, it is seen only in the Mediterranean region. The early farming successes of the *J* lineages spawned population booms and encouraged migration throughout much of the Mediterranean world. In fact, both haplogroup *J* and its subgroup *J2* are found at a combined frequency of around 30 percent among Jewish individuals.

HAPLOGROUP *J1*
Ancestral line: "Adam" → *M168* → *M89* → *M304* → *M267*

Haplogroup *J1* emerged during the Neolithic Revolution in the Middle East. Members of the *J1* clans shared the farming successes of the other *J* haplogroups. In particular some *J1* individuals moved back into North Africa and were quite successful, as evidenced by the fact that this haplogroup is currently seen there at its highest frequency. While other members of haplogroup *J1*, bearing the characteristic genetic marker *M267*, remained in the Middle East, some moved northward into western Europe, where today *J1* is found at low frequencies.

HAPLOGROUP *J2*
Ancestral line: "Adam" → *M168* → *M89* → *M304* → *M172*

The *M172* marker defines a major subset of haplogroup *J*, which arose from the *M89* lineage. Haplogroup *J2* is found today in North Africa, the Middle East, and southern Europe. In southern Italy it occurs at frequencies of 20 percent, and in southern Spain, 10 percent of the population carries this marker.

HAPLOGROUP K

Ancestral line: "Adam" → *M168* → *M89* → *M9*

The marker *M9* first appeared in a man born around 40,000 years ago in Iran or south-central Asia, which marked a new lineage diverging from the *M89* Middle Eastern clan. His descendants spent the next 30,000 years populating much of the planet.

This large lineage, called the Eurasian Clan, dispersed gradually over thousands of years. Seasoned hunters followed the herds ever eastward, along the vast superhighway of the Eurasian steppe. Eventually their path was blocked by the massive mountain ranges of south-central Asia: the Hindu Kush, the Tian Shan, and the Himalayas.

The three mountain ranges meet in the center of a region known as the Pamir Knot, located in present-day Tajikistan. Here the tribes of hunters split into two groups. Some moved north into central Asia, others moved south into what is now Pakistan and the Indian subcontinent. These different migration routes through the Pamir Knot region gave rise to more separate lineages. Most people native to the Northern Hemisphere trace their roots to the Eurasian Clan. Nearly all North Americans and East Asians are descended from this man, as are most Europeans and many Indians.

HAPLOGROUP K2

Ancestral line: "Adam" → *M168* → *M89* → *M9* → *M70*

Not all *M9* descendants challenged the problem of the Pamir Knot. Others stayed in the relatively fertile environment of the Near East. There, some 30,000 years ago the marker *M70* appeared and today defines this haplogroup, *K2*. Ancient members of haplogroup *K2* dispersed across the Mediterranean world. They traveled west along the coast of North Africa and also along the Mediterranean coastline of southern Europe. These movements suggest an intriguing possibility that the *M70* marker may have been carried by Mediterranean traders such as the Phoenicians. These seafaring people established a formidable, first-millennium B.C. trading empire that spread westward across the Mediterranean from its origins on the coast of modern Lebanon. *M70* is found today throughout the Mediterranean, but it shows its highest frequency (about 15 percent) in the Middle East and in northeast Africa. Members of this haplogroup are also found in southern Spain and France.

HAPLOGROUP *L*
Ancestral line: "Adam" → *M168* → *M89* → *M9* → *M20*

This part of the *M9* Eurasian clan migrated south once they reached the rugged and mountainous Pamir Knot region. The man who gave rise to marker *M20* was possibly born in India or the Middle East. His ancestors arrived in India around 30,000 years ago and represent the earliest significant settlement of India. For this reason haplogroup *L* is known as the Indian Clan. More than 50 percent of southern Indians carry marker *M20* and are members of haplogroup *L,* even though they were not the first people to reach India.

HAPLOGROUP *M*
Ancestral line: "Adam" → *M168* → *M89* → *M9* → *M4*

As the frigid temperatures of the ice age loosened their grip, *M4* ancestors headed to the coastline of Southeast Asia. The first man with the marker may have been born some 10,000 years ago, although geneticists are not certain of the exact date. People carrying this distinct genetic marker today live primarily in Melanesia, Indonesia, and to a lesser extent, in Micronesia.

The marker may have dispersed through the islands with the spread of rice agriculture in the region. The pattern of settlement and intense exploitation of a few plant species was similar and occurred at around the same time in the Middle East and what is now China. Archaeological sites in northern China show evidence of millet (a wheat-like grain) cultivation beginning around 7,000 years ago. Rice farming had reached Indonesia's islands of Borneo and Sumatra by 4,000 years ago.

Genetic evidence suggests that the seafaring peoples who crossed large bodies of water to reach the island archipelagos brought farming skills with them, rather than merely passing knowledge from group to group. Learning more about such rare lineages is a primary goal of the Genographic Project.

HAPLOGROUP N
Ancestral line: "Adam" → *M168* → *M89* → *M9* → *LLY22G*

A member of Eurasian Clan peoples who traveled north through the Pamir Knot region gave rise to the *LLY22G* marker, which defines haplogroup *N*. This man was probably born in Siberia within the last 10,000 years, and today his descendants can effectively trace a migration of Uralic-speaking peoples during the last several thousand years. This lineage has dispersed throughout the generations, and is now found in southern parts of Scandinavia as well as northern Asia.

The cultures of Uralic-speaking people are extremely diverse. Most of those in northern Europe have been hunters and herders; the Hungarians, in their earliest history, were horse nomads of the steppes. Many Russians from the far north are also members of haplogroup *N*.

The Saami, or Lapps, an indigenous people of northern Sweden, Norway, Finland, and Russia, traditionally supported themselves with hunting and fishing, their movements dictated by the reindeer herds. There may be as few as 85,000 Saami left today. Small, indigenous communities like this are being forced into mainstream society as industry reduces their habitat. Projects like the Genographic Project may be the last opportunity to capture their genetic data and learn more about haplogroup *N*.

HAPLOGROUP *O*
Ancestral line: "Adam" → *M168* → *M89* → *M9* → *M175*

A man carrying the marker *M175* was born around 35,000 years ago in Central or East Asia. This ancestor was part of the *M9* Eurasian clan that, encountering impassable mountain ranges, migrated to the north and east. These early Siberian hunters continued to travel east along the great steppes, gradually moving through southern Siberia. Some, perhaps taking advantage of the Dzhungarian Gap used thousands of years later by Genghis Khan to invade central Asia, made it into present-day China.

By the time haplogroup *O* ancestors arrived in China and East Asia, the last ice age was near its peak. Encroaching ice sheets and central Asia's enormous mountain ranges effectively corralled them in East Asia, and there they evolved in isolation over the millennia. Today, some 80 to 90 percent of all people living east of central Asia's great mountain ranges are members of haplogroup *O*, the East Asian Clan. The marker *M175* is virtually nonexistent in western Asia and Europe.

There were actually two waves of migration into this region. While *M175* populated the region from the north, another group approached from the south. Descendants of the first migration who left Africa may have reached East Asia by 50,000 years ago. Also common throughout northeast Asia, their lineage is found at a frequency of 50 percent in Mongolia.

HAPLOGROUP *O1A*

Ancestral line: "Adam" → *M168* → *M89* → *M9* → *M175* → *M119*

Haplogroup *O1a* is defined by a genetic marker called *M119*, which first appeared in South China or Southeast Asia about 30,000 years ago. *O1a* members subsequently dispersed throughout much of Southeast Asia, and their descendants remain numerous there today. Another group of *M119* carriers took a longer journey, eastward though Asia, all the way to Taiwan, where *O1a* appears in frequencies of around 50 percent in several aboriginal populations.

HAPLOGROUP *O2*

Ancestral line: "Adam" → *M168* → *M89* → *M9* → *P31*

Roughly 30,000 years ago, a man first displayed the genetic marker *P31,* which now defines haplogroup *O2*. This man lived in eastern Asia, perhaps in southern China, and his descendants spread south into Southeast Asia, east to Korea, and north to Japan. This distinctly Asian haplogroup is most common today in Southeast Asian nations like Malaysia and Thailand.

HAPLOGROUP *O3*

Ancestral line: "Adam" → *M168* → *M89* → *M9* → *M175* → *M122*

The ancestral man who gave rise to marker *M122* was probably born in China or southeast Asia. The widespread distribution of this man's descendants—more than half of Chinese men— strongly suggests that the spread of his descendants was closely tied to the spread of agriculture. Members of haplogroup *O3* may well be the descendants of China's first rice farmers. The development of rice agriculture in East Asia led to a large population expansion. Archaeological evidence for the spread of rice agriculture to Japan, Taiwan, and Southeast Asia parallels the genetic data and suggests that a unique population carrying this marker expanded and spread throughout the region.

HAPLOGROUP P

Ancestral line: "Adam" → *M168* → *M89* → *M9* → *M45*

M45 arose around 35,000 years ago in a man born in central Asia. He descended from the *M9* Eurasian clan that had moved to the north of the mountainous Hindu Kush and onto the game-rich steppes of present-day Kazakhstan, Uzbekistan, and southern Siberia. Although big game was plentiful, the environment on the Eurasian steppes became increasing hostile as the glaciers of the last ice age began to expand. The reduction in rainfall may have induced desert-like conditions on the southern steppes, forcing this group to follow the herds of game north. To exist in such harsh conditions, they learned to build portable shelters made of animal skins, to improve weaponry, and to adapt their hunting techniques for use against the much larger animals encountered in colder climates. They compensated for the lack of stone by developing smaller points and blades—microliths—that could be mounted to bone or wood handles. Their tool kits also included bone needles for sewing animal skin clothing that would insulate them and allow them the range of motion needed to hunt the reindeer and mammoth. This clan's resourcefulness and ability to adapt was critical to survival in Siberia during the last ice age, a region where no other hominid species are known to have lived. The *M45* central Asian clan gave rise to many more; this man was the common ancestor of most Europeans and nearly all Native American men.

HAPLOGROUP Q

Ancestral line: "Adam" → *M168* → *M89* → *M9* → *M45* → *M242*

M242 arose some 15,000 to 20,000 years ago, with a man born in the savagely cold climate of Siberia. His descendants became the first explorers of North America. Despite frigid temperatures, some of the Siberian Clan gradually crossed Siberia's ice-free tundra to eastern Siberia. Once they reached the northeastern edge of Asia, they were ideally poised to enter a new world.

About 15,000 years ago they did just that. With much of Earth's water locked up in ice sheets, sea levels were some 330 feet (100 meters) lower than they are at present. Consequently, a landmass called Beringia connected present-day Siberia and Alaska and provided a crossing point for the peopling of the Americas.

There has been some debate about whether humans reached North or South America much earlier, prior to 20,000 years ago. However, the genetic data coincide with archaeological evidence for a Beringia crossing and lend support to the theory that this migration occurred about 15,000 years ago.

Members of haplogroup *Q* continued their migration farther south through the Americas. Just how they gained passage through the era's prevalent ice cover is unknown. Recent climate and geological evidence suggests that an ice-free Rocky Mountain corridor opened, allowing safe travel. A coastal migration route is another possible route. While some of the Siberian Clan remained in Asia (marker *M242* can be found in India and China as well as Siberia), almost all Native Americans are descendants of these people.

HAPLOGROUP *Q3*
Ancestral line: "Adam" → *M168* → *M89* → *M9* → *M45* → *M242* → *M3*

Soon after the first explorers reached the Americas a new marker, *M3,* arose. This ancestor, born in North America

10,000 to 15,000 years ago, is the patriarch of the most wide-spread lineage in the Americas. Nearly all native South Americans and most native North Americans are descended from this line.

While descendants of the Siberian Clan, the first wave of migrants to reach North America, can be found in both Asia and America, $Q3$ descendants are found only in America. Because of this, geneticists believe that their common ancestor was born after the Bering Strait land bridge was once again submerged around 10,000 to 15,000 years ago.

Whether traveling an ice-free corridor along the Rocky Mountains, or down the coastline, this group continued to move south. Within a thousands years, they had reached the tip of South America. After the rigors of life dominated by freezing temperatures, the explorers now found land abundant with food and natural resources. Although as few as 10 to 20 people may have initially made the crossing from Siberia to North America, the abundance of game led to a huge increase in human populations.

HAPLOGROUP R

Ancestral line: "Adam" → $M168$ → $M89$ → $M9$ → $M45$ → $M207$

After spending considerable time in central Asia, refining skills to survive in harsh new conditions and exploit new resources, a group from the central Asian Clan began to head west toward the European subcontinent.

An individual in this clan carried the new $M207$ mutation on his Y chromosome. His descendants ultimately split into two distinct groups, with one group continuing westward onto the European subcontinent and the other turning south to ultimately end its journey in the Indian subcontinent. This genetic distribution supports both the linguistic and archaeological evidence suggesting that a large migration from the Asian steppes into India occurred within the last 10,000 years. However, the ancient migrations of this second group, and the distribution of genetic lineages to which it ultimately gave rise, still remain largely a mystery because there is little recorded data to work with.

HAPLOGROUP R1

Ancestral line: "Adam" → *M168* → *M89* → *M9* → *M45* → *M207* → *M173*

Members of haplogroup *R* are descendants of Europe's first large-scale human settlers. Their lineage is defined by Y-chromosome marker *M173*, which shares a westward journey of *M207*-carrying, central Asian steppe hunters. The descendants of *M173* arrived in Europe around 35,000 years ago and immediately began to make their own dramatic mark on the continent. Soon after their arrival, the era of the Neandertals came to a close. Smarter, more resourceful human descendants of *M173* likely outcompeted Neandertals for scarce ice age resources and brought on their ultimate demise.

The long journey of this lineage was further shaped by the preponderance of ice at this time. Humans were forced to southern refugia in Spain, Italy, and the Balkans. Years later, as the ice retreated, they moved northward out of these isolated refugia and left an enduring, concentrated trail of the *M173* marker in their wake. Today, for example, the marker's frequency remains very high in Spain and the British Isles where it was carried by *M173* descendants who had weathered the last ice age in the Iberian refugium.

HAPLOGROUP R1A1

Ancestral line: "Adam" → *M168* → *M89* → *M9* → *M45* → *M207* → *M173* → *M17*

Sometime between 10,000 to 15,000 years ago, a man of European origin was born in present-day Ukraine or southern Russia. His nomadic descendants would eventually carry his genetic marker, *M17*, from the steppes to places as far away as India and Iceland. Archaeologists speculate that these people were the first to domesticate the horse, which would have eased their distant migrations.

In addition to genetic and archaeological evidence, the spread of languages can trace prehistoric migration patterns of the *R1a1* clan, whose descendants may be have spread the Indo-European languages. The world's most widely spoken language family, Indo-European tongues include English, French, German, Russian, Spanish, several Indian languages such as Bengali and Hindi, and numerous others. Many of the Indo-European languages share similar words for animals, plants, tools, and weapons.

Some linguists believe that the Kurgan people, nomadic horsemen roaming the steppes of southern Russia and the Ukraine, were the first to speak and spread a proto-Indo-European language, some 5,000 to 10,000 years ago. Genetic data and the distribution of Indo-European speakers suggest the Kurgan, named after their distinctive burial mounds, may have been descendants of *M17*.

Today a large concentration—around 40 percent—of the men living from the Czech Republic across the steppes to Siberia, and south throughout central Asia are descendants of this clan. In India, around 35 percent of the men in Hindi-speaking populations carry the *M17* marker, whereas the frequency in neighboring communities of Dravidian speakers is only about 10 percent. This distribution adds weight to linguistic and archaeological evidence suggesting that a large migration from the Asian steppes into India occurred within the last 10,000 years.

The *M17* marker is found in only 5 to 10 percent of Middle Eastern men. This is true even in Iranian populations where Farsi, a major Indo-European language, is spoken. Despite the low frequency, the distribution of men carrying the *M17* marker in Iran provides a striking example of how climate conditions, the spread of language, and the ability to identify specific markers can combine to reveal migration patterns of individual genetic lineages. In the western part of the country, descendants of the Indo-European Clan are few, encompassing perhaps 5 to 10 percent of the men. However, on the eastern side, around 35 percent of the men carry the *M17* marker. This distribution suggests that the great Iranian deserts presented a formidable barrier and prevented much interaction between the two groups.

HAPLOGROUP *R1B*
Ancestral line: "Adam" → *M168* → *M89* → *M9* → *M207* → *M173* → *M343*

Around 30,000 years ago, a descendant of the clan making its way into Europe gave rise to marker *M343*, the defining marker of haplogroup *R1b*. These travelers are direct descendants of the people who dominated the human expansion into Europe, the Cro-Magnon. The Cro-Magnon created the famous cave paintings found in southern France, providing archaeological evidence of a blossoming of artistic skills as people moved into Europe. Prior to this, artistic endeavors were mostly comprised of jewelry made of shell, bone, and ivory; primitive musical instruments; and stone carvings. The cave paintings depict animals and natural events important to Paleolithic life such as spring molting, hunting, and pregnancy. The paintings are far more intricate, detailed, and colorful than anything seen prior to this period.

HAPLOGROUP *R2*
Ancestral line: "Adam" → *M168* → *M89* → *M9* → *M207* → *M124*

About 25,000 years ago, a man from southern central Asia first displayed the genetic marker *M124*. His descendants migrated to inhabit what is now Pakistan and also farther east in modern India. Today they are found in Northern India, Pakistan, and southern central Asia at frequencies of 5 to 10 percent. The *R2* lineage also belonged to the second major human migration into India, long after the first large wave of African migrants 50,000 to 60,000 years ago.

Members of the haplogroup *R2* are also found in Eastern Europe among the Gypsy populations. Their genes tie these wandering peoples back to their origins on the Indian subcontinent. These ancient migrations and the distribution of their genetic lineages still remain mysterious as scientists search for more data with which to uncover the history of this haplogroup.

GLOSSARY

BASE

The chemical building blocks of DNA. Abbreviated A, T, C, and G (for adenine, cytosine, thymine, and guanine), these bases pair up to form the "stairs" of the DNA double helix and always combine in the same patterns: A with T and C with G.

CELL

The smallest unit of living matter that can operate independently.

CHROMOSOME

Long strands of DNA on which genes are found. Each human cell has 46 chromosomes in 23 pairs. One member of each pair is inherited from the mother, the other from the father.

DARWIN, CHARLES

His work became the foundation of modern evolutionary theory. Charles Darwin's 1859 book *The Origin of Species* promoted a theory of evolution by natural selection and challenged Victorian-era ideas about the role of humans in the universe. Darwin's theories were based on a constantly evolving natural world and held that each generation of a species had to compete for survival. Survivors held some natural advantages over their unfortunate relatives and passed those character-istics on to their progeny, thus overrepresenting these favored genetic types in the next generation. Darwin also advanced the idea that species were descended from a common ancestor. Darwin's work became the foundation of modern evolutionary theory.

DNA (DEOXYRIBONUCLEIC ACID)
The double helix-shaped molecule that holds an organism's genetic information. DNA is composed of sugars, phosphates, and four nucleotide bases: adenine, guanine, cytosine, and thymine (A, G, C, and T). The bases bind together in specific pairs (see Base).

DOUBLE HELIX
The shape of DNA, much like a spiral staircase or twisted ladder. The stairway's railings are composed of sugars and phosphates. Its sides contain the patterned base pairs: A, T, C, and G. When a cell divides for reproduction, the helix unwinds and splits down the middle like a zipper in order to copy itself.

GENES
Segments of DNA that are the basic functional units of heredity. Genes are determined by an ordered sequence of chemical bases found in a unique position on a specific chromosome. Their "blue-print" guides protein production, which determines how different cells in the body function. Inherited genes also control an animal's unique set of physical traits.

GENETIC MARKER
Random mutations in the DNA sequence which act as genetic mile-stones. Once markers have been identified they can be traced back in time to their origin—the most recent common ancestor of everyone who carries the marker.

GENOME
The total DNA sequence that serves as an instruction manual for all proteins created in our body. Two copies of the genome are found inside each of our cells.

HAPLOGROUP
Branches on the tree of early human migrations and genetic evolution. Haplogroups are defined by genetic mutations or "markers" found in Y-

chromosome and mtDNA testing. These markers link the members of a haplogroup back to the marker's first appearance in the group's most recent common ancestor. Haplogroups often have a geographic relation.

HAPLOTYPE
A person's individual footprint of all tested genetic markers. Even the difference of a single genetic marker delineates a distinct haplotype.

HEREDITY
The total sum of genetic information that humans pass on from generation to generation.

MELANIN
Melanin, the skin's brown pigment, is a natural sunscreen that protects tropical peoples from the many harmful effects of ultraviolet (UV) rays. But when UV rays penetrate the skin they also produce beneficial vitamin D, so some exposure to them is necessary. This delicate balancing act explains why the peoples that migrated to darker, colder climes also developed lighter skin color. As people moved to areas with lower UV levels, their skin lightened so that UV rays could penetrate and produce essential vitamin D. In some cases a third factor intervened. Coastal peoples who eat diets rich in seafood enjoy an alternate source of vitamin D. That means that some Arctic peoples, for example, can afford to remain dark-skinned even in low UV climes.

MITOCHONDRIA
A remnant of an ancient parasitic bacteria that now helps to produce energy inside the cell. A mitochondrion has its own genome, present in only one copy, which does not recombine in reproduction. This genetic consistency makes mitochondrial DNA a very important tool in tracking genetic histories.

MITOCHONDRIAL DNA (mtDNA)
Genetic material found in the mitochondria. It is passed from females to their offspring without recombining, and thus is an important tool for geneticists.

NUCLEOTIDE
A DNA building block that contains a base, or half of a "staircase step," and sugars and phosphates which form the "railing." Nucleotides join together to form DNA's distinctive double helix shape.

NUCLEUS
The part of the cell in which chromosomes reside.

PHYLOGENY
The evolutionary development of a species. Phylogeny is sometimes represented as a tree that shows the natural relations and development of all species.

POPULATION GENETICS
The study of genetic variation in a species.

PROTEINS
Linear sequences of amino acids that are the building blocks of cells. Each protein has a specific function that is determined by the "blueprint" stored in DNA.

RECOMBINATION
The process by which each parent contributes half of an offspring's DNA, creating an entirely new genetic identity. This process mixes genetic signals, so that nonrecombining DNA, passed intact through the generations, is most important to population genetics.

REPLICATION
The process by which two DNA strands separate, with each helping to duplicate a new strand. During reproduction, the DNA double helix unwinds and duplicates itself to pass on genetic information to the next generation. Because bases always form established pairs (AT and CG), the sequence of bases on each strand will attract a corresponding match of new bases. Only occasional errors occur—about one for every billion base-pair replications.

SEQUENCING
Determines the order of nucleotides for any particular DNA segment or gene. The order of a DNA string's base pairs determines which proteins are produced, and thus the function of a particular cell.

SEXUAL SELECTION
Special form of natural selection based on an organism's ability to mate. Some animals possess characteristics that are more attractive to potential mates, such as the distinctive plumage of some male birds. Individuals with such characteristics mate at higher rates than those

without, ensuring more next generation offspring will inherit the desirable trait. As generations procreate the desirable trait becomes increasingly common, further boosting the sexual disadvantage for individuals who lack the desired trait. The effect can be particularly dramatic when one individual controls mating with a large number of potential partners.

SINGLE NUCLEOTIDE POLYMORPHISM
Small, infrequent changes that help to create an individual's own unique DNA pattern. When a single nucleotide (A, T, G, or C) is altered during DNA replication, due to a tiny "spelling mistake," the genome sequence is altered.

TRAIT
Physical characteristics, like eye color or nose shape, which are determined by inherited genes.

X AND Y CHROMOSOMES
The two chromosomes that determine sex. Females have two X chromosomes while males have one X and one Y. When chromosomes pair, the mismatched Y determines male gender. Because of the mismatch, part of the Y chromosome does not recombine with the X during reproduction. The nonrecombining part of the Y chromosome contains a sequence of DNA passed intact from males to their sons through the generations, giving population geneticists a useful tool for studying human history.

FURTHER READING

The literature devoted to human origins and early migratory history is vast, ranging from popular best sellers to technical volumes on genetics. The books listed below are recommended for readers interested in additional sources of information.

BOOKS
Cavalli-Sforza, Luigi Luca. *Genes, Peoples, and Languages*. Berkeley, CA: University of California Press, 2001.
A good overview of the work of Cavalli-Sforza and his colleagues over the past half-century.

Fagan, Brian. *People of the Earth: An Introduction to World Prehistory*, 11th ed. East Rutherford, NJ: Prentice Hall, 2003.
An excellent overview of the human paleontological and archaeological record.

Jobling, Mark, Matthew Hurles, and Chris Tyler-Smith. *Human Evolutionary Genetics: Origins, People and Disease*. New York: Garland Publishing, Inc., 2003.
A fantastic introductory textbook with lots of scientific detail.

Jones, Stephen, ed., et al. *The Cambridge Encyclopedia of Human Evolution*. New York: Cambridge University Press, 1992.
A wonderful overview of all aspects of human evolution, from physical anthropology to recent human demographic trends.

Klein, Richard and Blake Edgar. *The Dawn of Human Culture*. Hoboken, NJ: John Wiley & Sons, Inc., 2002.
Focuses on the transition to modern behavior in Africa around 50,000 years ago.

Oppenheimer, Stephen. *The Real Eve: Modern Man's Journey Out of Africa*. New York: Carroll & Graf, 2004.
A good overview of Y-chromosome and mtDNA evidence for human migratory patterns.

Smolenyak, Megan and Ann Turner. *Trace Your Roots With DNA: Using Genetic Tests to Explore Your Family Tree*. Emmaus, PA: Rodale Press, 2004.
Excellent coverage of "genetic genealogy," explaining how to make sense of the variety of commercially available DNA tests.

Sykes, Brian. *The Seven Daughters of Eve*. New York: W. W. Norton & Company, 2002.
A best seller focusing on the seven founding mtDNA lineages in Europe. A good source for the basics of mtDNA migration patterns in western Eurasia.

Wells, Spencer. *The Journey of Man: A Genetic Odyssey*. New York: Random House, 2004.
A slightly more technical version of the story told in *Deep Ancestry*, with particular focus on the Y chromosome and its use in tracing human migrations over the past 50,000 years.

WEB SITES
In addition, several Web sites offer good information on this topic:

The Genographic Project
http://www.nationalgeographic.com/genographic
Gives information on the Project, its organization, goals of the scientific research, and provides a narrative of human migrations over the past 200,000 years.

Other Web sites focusing on anthropology include the following:

Anthropology in the News
http://anthropology.tamu.edu/news.htm
From Texas A&M University, an up-to-date compilation of news stories from the field of anthropology.

John Hawks' Anthropology Weblog: Paleoanthropology, Genetics, and Evolution
http://johnhawks.net/weblog/
A blog devoted to topical issues in anthropology.

Dienekes' Anthropology Blog
http://dienekes.blogspot.com
A blog with a strong focus on genetic studies.

Genealogy DNA Listserv Archives
http://archiver.rootsweb.com/th/index/GENEALOGY-DNA/
An internet forum for genetic genealogy, with many posts focused on deep ancestry.

JOURNALS
Major journals publishing scientific papers in the field of genetic anthropology include the following:

American Journal of Human Genetics
http://www.journals.uchicago.edu/AJHG/

Nature Genetics
http://www.nature.com/ng/

Nature
http://www.nature.com/nature/

Science
http://www.sciencemag.org

ABOUT THE AUTHOR

D r. Spencer Wells is a 37-year-old geneticist, anthropologist, writer, and filmmaker. A child prodigy with a love for both history and science, Wells spent a great deal of time in the laboratory with his mother as she pursued a doctorate in biology. It was there that he realized how scientific methods could be used to explore questions of the past. Entering the University of Texas at age 16, Wells majored in biology and graduated Phi Beta Kappa in 1988.

In 1989 Wells began his doctoral work under evolutionary geneticist Richard Lewontin at Harvard University. Wells devoted his time to understanding how tiny changes in DNA could reveal information about patterns of human evolution and migration. After completing his Ph.D., Wells headed to Stanford University in California to work with population geneticist Luigi Luca Cavalli-Sforza. There, Wells's work focused on using the "documents" stored in the DNA of living people to study human origins and how our ancestors came to populate the world. In a truly multidisciplinary effort, Wells combined his findings with the work of linguists, palaeoclimatologists, archaeologists, and geneticists to gain a more complete picture of how modern human beings came to be so widely scattered about the planet.

After leaving Stanford, Wells embarked to the ex-Soviet republics of the Caucasus and Central Asia to study isolated, native populations. He collected more than 2,000 specimens for DNA analysis. Then, Wells headed a research group at Oxford University's Wellcome Trust Centre for Human Genetics; in 2000, Wells was appointed Director of Population Genetics for a biotech company in Cambridge, Massachusetts. Soon after, he began work on his next project: the PBS/National Geographic 2003 documentary *The Journey of Man: A Genetic Odyssey* and the book of the same name. Filming took him to the farthest reaches of the globe in search of indigenous human populations—like the Kalahari Bushmen, the Chukchi reindeer herders of the Russian Arctic, the Australian Aborigines, and Native Americans—who hold more hidden human history in their DNA. In 2005, Wells became a National Geographic Society Explorer-in-Residence and Director of the Genographic Project, the largest genetic study of human migration ever mounted.

Writer's Acknowledgments

A project the size of Genographic wouldn't be possible without the dedicated work of many people. Throughout the two years of planning and during the first year since the Project's launch in April of 2005, the National Geographic "Geno team" members have proven themselves to be an amazingly dedicated crew. Alex Moen, Lucie McNeil, Jodi Barr, Anne Gregal, Glynnis Breen, Sarah Laskin, Kim McKay, Frank Vidergar, Carol Young, and Merone Demissie have done a huge amount of behind-the-scenes work to keep the project going. Likewise, David Yaun, Kris Lichter, and the team at IBM, including Ajay Royyuru's group at the Watson Labs in Yorktown Heights, have been invaluable in keeping the complexities of the Project under control on the computational front. My right-hand man, Jason Blue-Smith, has done a wonderful job juggling everything from contract negotiations to buffer solutions for fieldwork to the illustrations for this book—thanks, Jase. The Digital Media team at National Geographic and Terra Incognita have done a great job creating an award-winning Web site. Finally, I'd like to thank the senior management at National Geographic and IBM, especially Terry Garcia, John Fahey, and Nick Donofrio, and Ted Waitt at the Waitt Family Foundation, for believing in this project during the planning stages, and for their continuing support now that it has launched.

ILLUSTRATION
CREDITS

Cover: Hillman and Partners. Back Cover: David Evans. Interior: 32, From the Harry H. Laughlin Papers, Truman State University, Courtesy Dolan DNA Learning Center; 35, Scala/Art Resource, NY; 66, Jenny Kubo, NGS: 94, Bryan & Cherry Alexander Photography; 110, John Gurche; 119, Bradshaw Foundation, Geneva/www.bradshawfoundation.com; 136 (top row), David Brill; 136 (lower left), Giraudon/Art Resource, NY; 136 (lower center), Réunion des Musées Nationaux/Art Resource, NY; 136 (lower right), Bridgeman Art Library/Getty Images; 152 (upper left), José Azel/AURORA; 152 (upper right), James L. Stanfield; 152 (lower left), Wendy Stone/CORBIS; 152 (lower right), Joy Tessman/NGS Image Collection.

All maps and diagrams were created by the National Geographic Society, with thanks to Justin Morrill at The M Factory, Inc. Some figures were based on data from the following publications:

Chapter 1.3: L. L. Cavalli-Sforza, I. Barrai, and A. W. Edwards. "Analysis of Human Evolution under Random Genetic Drift." *Cold Spring Harbor Symposia on Quantitative Biology* 29 (1964): 9-20.

Chapter 2.1: G. A. Harrison et al. *Human Biology.* New York: Oxford University Press, 1988.

Chapter 3.2: M. A. Jobling, M. E. Hurles, and C. Tyler-Smith. *Human Evolutionary Genetics: Origins, Peoples, and Disease.* New York: Garland Publishing, 2004.

Chapter 3.4: L. L. Cavalli-Sforza, P. Menozzi, and A. Piazza. *The History and Geography of Human Genes.* Princeton, NJ: Princeton University Press, 1994.

Chapter 3.6: M. Richards et al. "Tracing European Founder Lineages in the Near Eastern mtDNA Pool." *American Journal of Human Genetics* 67 (2000): 1251-76.

Chapter 5.2: Christopher R. Scotese, PALEOMAP Project.

INDEX

Boldface indicates
illustrations.

A haplogroup: mtDNA 188;
Y-chromosome 143, 146,
204–205
Aborigines 121, 126
Adam (common ancestor):
African 158–160; Eurasian
132
Adenine 14
Africa: coastal migration
from 123–126; common
ancestors 156–159;
hominids 107; *Homo
erectus* 116; map of
coastal migration 125;
map of major mitochon-
drial haplogroups 151;
map of major Y chromo-
some haplogroups 147; as
most diverse continent
150; oldest evidence
for ancestors 116; rock
art 118; skin color
153–155
Africans *(afer)* 17, 18
Agriculture 64–68, 82;
map 69

Americans *(americanus)* 17
Americas: haplogroup *Q*
86–87; mtDNA lineages
104
Ammerman, Albert 67, 71
Ancestry: pull of 10–11
Animal kingdom 57
Anthropology: revolution in 25
Archaea 57
Archaeology 62–63
Archaic *Homo sapiens* 107
Artifacts 117; stone **136**
Asia: agriculture 66; genetic
linkages in population
100; haplogroup *Q*
86–87; haplogroup *R1a*
77; ice age 92, 111–112;
mtDNA lineages 104;
rock art 118; Siberia
92–96, 99–100; *see also*
Central Asia; East Asia;
Eurasia
Asians *(asiaticus)* 17
Aspen trees 87–88
Australia: coastal migration
from Africa to 123–126;
early evidence for
humans/tools 116, **136**;
hominids 117; joined to

New Guinea 121–122;
migration to 117–123;
rock art 118, 120
Australians 17
Australoids 17

B haplogroup: mtDNA
189–190; Y-chromosome
142, 143, 146, 205
Bacterial DNA 58
Balanovska, Elena 168
Balkans: haplogroups 52, 54;
ice age 80, 81, 111
Bantu **152**
Bantu language speakers
145–146, 150
Bases, DNA 14; defined 229
Bedouins 59
Bertranpetit, Jaume 166
Binomial system of nomen-
clature 16–17, 57–58
Biological data: ways of
looking at 19–20
Biotechnology revolution
25
Bipedalism 137
Blair, Nebraska 55
Blood groups 19; polymor-
phism studies 22–24

JOIN US ON A LANDMARK STUDY OF THE HUMAN JOURNEY.

You can participate in this real-time scientific study by purchasing a Genographic Project Public Participation Kit. Your results will reveal your deep ancestry along a single line of direct descent (paternal or maternal) and trace the migration routes your ancestors followed thousands of years ago.

Proceeds from the sale of kits benefit the Genographic Legacy Fund, which will fund projects seeking to strengthen indigenous and traditional communities through education initiatives, cultural conservation efforts, and indigenous language preservation and revitalization programs.

The Genographic Project Public Participation Kit costs U.S. $99.95 (plus shipping and handling and tax if applicable). The purchase price also includes the cost of the testing and analysis, which will take place once your sample is submitted to the lab. Orders should be placed via our Web site. Orders cannot be taken over the phone.

THE KIT INCLUDES:

1. DVD with a Genographic Project overview hosted by Dr. Spencer Wells, visual instructions on how to collect your DNA sample, and a bonus feature program: the National Geographic Channel/PBS production *The Journey of Man*.
2. Exclusive National Geographic map illustrating human migratory history and created especially for the launch of the Genographic Project.
3. Swab kit, instructions, and a self-addressed envelope in which to return your sample.
4. Detailed brochure about the Genographic Project, featuring stunning National Geographic photography.
5. Confidential Genographic Participant ID (GPID) to access your results anonymously at *www.nationalgeographic.com/genographic*.

To learn more about the Project and how to participate, please visit our Web site at *www.nationalgeographic.com/genographic*

ORDER YOUR PARTICIPATION KIT TODAY!

A research partnership of National Geographic and IBM

THE WAITT FAMILY FOUNDATION.

Global field science supported by the Waitt Family Foundation